TROPICAL *fRUIT*

GLENN
TANKARD

TROPICAL
fruit

*An Australian guide
to growing and using
exotic fruits*

VIKING

Viking
Penguin Books Australia Ltd
487 Maroondah Highway, PO Box 257
Ringwood, Victoria 3134, Australia
Penguin Books Ltd
Harmondsworth, Middlesex, England
Viking Penguin, A Division of Penguin Books USA Inc.
375 Hudson Street, New York, New York 10014, USA
Penguin Books Canada Limited
10 Alcorn Avenue, Toronto, Ontario, Canada M4V 3B2
Penguin Books (N.Z.) Ltd
182–190 Wairau Road, Auckland 10, New Zealand

First published by Thomas Nelson Australia 1987
This edition published by Penguin Books Australia Ltd 1990

10 9 8 7 6 5 4 3

Typeset by Bookset
Designed by Tom Kurema
Printed and bound through Bookbuilders Limited, Hong Kong

National Library of Australia
Cataloguing-in-Publication data

Tankard, G. J. (Glenn J.).

Tropical fruit.

Bibliography.
Includes index.
ISBN 0 670 90205 5.

1. Fruit-culture – Australia. 2. Tropical fruit –
Australia 3. Cookery (Fruit). I. Taranto, Jenny. II.
Title. III. Title: Exotic tree fruit for the Australian
home garden.

634.0994

CONTENTS

This book is dedicated to the rare fruit explorers and pioneers, who have awakened millions of sleepy Aussie taste-buds with new and unusual taste sensations.

'Strange taste of a tropical fruit,
Romantic language of the Portuguese,
Melody on a wooden flute,
Summer floating in the summer breeze.'
Only a dream in Rio, James Taylor.

ACKNOWLEDGEMENTS

Tropical Fruit would not have been written without the help of many people connected with the rare fruit industry in Australia. My special thanks and appreciation go to the following people for their assistance; David Chandlee and Lauren Gartrell, John and Jacky Marshall, Bruno and Deidre Scomazzon, Alan and Suzie Carle, Don and Christine Gray, Diane Cilento, Ian and Dawn Wilson, Joe and Ivy Zappala, David Higham, Malcolm Coates, Brian Watson, Hugh Skinner, Paul Recher, John Vanderbyl, David Wallace, Danny Latimer, Gordon Vallance, Elliot Tuckwell, Greg Morton, Danny Callaghan, Nick and Carol Gilbert, Zam Chaudri and Don Maggs. The chapter on rare fruit discoveries in Malaysian Borneo was researched, compiled and written by rare fruit explorers David Chandlee and Lauren Gartrell (Treefarm, El Arish, Qld). Editorial assistance was generously provided by John and Jacky Marshall (Cairns, Qld). Photo contributions were made by David Chandlee (El Arish, Qld), Diane Cilento (Mossman, Qld), in co-operation with the Queensland Tourist and Travel Corporation (Brisbane, Qld), Russell Francis (Cairns, Qld), Peer Productions (Cairns, Qld), Bruno Scomazzon (Mossman, Qld), Tom Hoult of Macadamia Plantations of Australia Nursery (Dunoon, NSW), Jim Truscott (Cairns, Qld). Some extra special thanks to Jenny Taranto for her lovely fruit sketches and to my dear sister Sarah for helping out with all the typing.

INTRODUCTION

INTRODUCTION

They journey to remote and exotic destinations in search of a strange new taste, a strange new fruit. From the upper reaches of the Amazon River in tropical Peru and Brazil to the Incan territories high up in the Andes mountains. From the rainforests of the Western Ghats in India to the jungles of Papua New Guinea, Sarawak and Sabah. From the swamps of Thailand to the island paradise of Bali. They race against time, progress and the continual destruction of the wilderness areas of the world. When they return it is often with some exciting new fruits and fascinating tales of their adventures in these distant lands. They are the plant hunters and seed collectors . . . their search for rare and exotic fruit continues.

Some fruits have been cultivated for thousands of years and have played an important part in the diet of the local people. The extent to which they were loved and cherished by these people is reflected in their art, folklore, and religious ceremonies. Images of the fruit are often depicted on clay pots and woven cloth. There are many lesser known fruits with promise that still remain virtually unknown. Their turn will come, and they too will be appreciated for the refreshing new tastes they bring us.

These days all we have to do is stroll into our local supermarket to find some colourful new fruit on the shelves. They come to us in all sorts of unusual shapes and sizes. Many are quite attractive, enticing us with their good looks and the promise of exquisite delights within. Others don't look quite so inviting, and are covered in sharp, stout spines, lumps and bumps, or even long, fleshy hairs! Our curiosity may get the better of us, and we cautiously taste the fruit for the first time. Often, much to our surprise, the most unusual-looking fruits turn out to be the most delicious, so don't let appearances deceive you. Some fruits have a taste that you'll probably enjoy instantly, others are more of an acquired taste and take a little longer to appreciate. Apart from their pleasant eating qualities, the fruits also play a useful part in our diets. Some are very nutritious and represent high energy foods, while others contain important vitamins and minerals necessary to keep our bodies running smoothly.

With such a fabulous range of new fruit trees becoming available, there's no doubt that many exciting, challenging and rewarding days lie ahead of us in our home gardens. Just like the adventurous rare fruit explorers, we too can now learn some of the wonderful secrets of these exotic delights.

WHAT FRUITS SHALL WE GROW?

What a decision to have to make when there are so many tasty fruits to choose from. Maybe some juicy mangos, some 'lemon meringue pie' rollinias, or some of those fabulous Amazon abius. What about casimiroas, cherimoyas, mangosteens, and caimitos? We certainly can't forget the lychees, and how about those delicious, hairy-looking rambutans? Wouldn't it be great if we could grow them all. Unfortunately most home gardens just aren't big enough. Perhaps we could get together with our neighbours, and each be responsible for planting a few different trees, and when harvest time comes we could carry out some friendly bartering. If the trees are growing well, then there's bound to be plenty of fruit to share.

After we've made up a list of our favourite fruits all we have to do is race out and plant them, right? Not so fast! Every fruit tree has its own special 'recipe' for growing healthily and bearing lots of luscious fruit. The 'microclimate' in our home garden has a lot to do with a plant's chances of success or failure. 'Microclimate' refers to the environmental conditions that exist in our home garden, and involves the interaction of climate, soils and any other factors likely to affect plant growth. The home garden microclimate can differ substantially from suburb to suburb, and even from house to house in the same street.

Many of the rare and exotic fruit trees mentioned in this book are native to the warmer climates of the world where frosts and low temperatures seldom occur. Often they are found growing wild in lush tropical and subtropical rainforests where they are sheltered from cold winds and thrive in the warm, humid atmosphere of the jungles. Other trees are more hardy and come from the hot, dry desert and savannah lands, or the cooler temperate zone where sub-zero temperatures are not uncommon. Some trees actually originate in tropical regions close to the equator, however they are found growing at such high elevations up in the mountains that the climate they experience is sub-tropical, and sometimes even temperate. Our own home gardens are very remote from these native tree habitats and often the environment they encounter here is totally alien to them.

We should carefully consider each tree's chances of survival before we rush out and plant. It is a good idea to study our growing site first, then out of a range of fruit trees that we fancy, select the ones we consider will be most suitable. The growing conditions don't necessarily have to be perfect, just so long as our trees are healthy and bear fruit regularly, that's the main aim. With a little extra care and attention many tropical fruit trees can be grown in cooler areas than they are accustomed to, so don't be afraid to give them a good try. (Refer to Modifying the Home Garden Microclimate.)

SELECTING THE BEST GROWING SITE IN YOUR GARDEN

Before we begin planting, we should take a close look at the growing conditions in our home garden. This will help us to locate the best positions for planting. Is there one spot in your garden that is warmer and more sheltered than another? Is the soil deep and well drained there? Most exotic fruit trees grow best in warm, sunny, frost-free positions, sheltered from damaging winds. Soils should be well structured and free draining.

Some plants are hardier than others and will grow fruit satisfactorily in less favourable positions than some of the more sensitive species. Armed with a sound knowledge of our local growing conditions, we should be better equipped to select the most appropriate planting position for each tree. A happy, healthy fruit tree will provide us with many a nutritious and mouth-watering harvest.

1. GOOD SOIL STRUCTURE

Most fruit trees grow best in deep, free-draining soils where their roots can spread freely without restriction. Waterlogged soils can result in oxygen deficiency, mineral toxicity, and provide a perfect breeding ground for the fungal organisms that cause root disease. Every effort should be made to select a growing spot with good soil drainage. Avoid positions where there is likely to be excessive water runoff from nearby garden areas. Sites where the water table is high and springs and seepages are evident should also be avoided. You can carry out your own soil depth test by digging a hole about a metre or more deep. Avoid plantings in areas with impervious clay bands, hardpans, or rock formations occurring too close to the soil surface. This is particularly advisable if avocados are to be

grown, as they are very susceptible to the root rot fungus, *Phytophthora cinnamomi*, which thrives in waterlogged soils. Conversely, soils that are too well drained can also lead to problems. Some very sandy soils tend to dry out rapidly and may lead to water stressing of trees during long, hot-dry periods. These soils require extensive watering during the growing months and this can be a nuisance if you plan to spend your vacations elsewhere. If you are unable to locate a suitable spot anywhere in your garden, don't despair, there are ways to enable you to improve soil structure artificially. (Refer to Modifying the Home Garden Microclimate.)

2. FREEDOM FROM DAMAGING WINDS

Most fruit trees require protection from strong winds. These winds can dessicate new growth, break tree branches, and damage developing fruits by causing them to rub against tree stems. Tree training is also made difficult as the plants tend to develop a lop-sided slant as they grow away from the wind. Shallow rooted trees such as the papaya and babaco can be completely blown over, especially when fully laden with fruit.

Cold winds, particularly when they occur unseasonally and last for extended periods, can be just as damaging as sub-zero temperatures. They cause tree and leaf damage and reduce flower initiation. They can also retard tree growth, in some cases resulting in a serious decline in general health. Hot, drying winds can also be a problem.

In seaside locations, salt-laden winds have been known to cause more damage to tree foliage, through salt spray, than from the force of the wind itself. If the only growing site available to you is frequented by damaging winds then some extra effort will be necessary in order to provide increased shelter. (Refer to Modifying the Home Garden Microclimate.)

3. FREEDOM FROM FROSTS

Most mature fruit trees usually tolerate light frosts and only suffer minor damage to branches and leaves. Young trees are generally not so fortunate and are very vulnerable for the first few years of growth. The more severe frosts are likely to kill or severely damage plants of all ages. Some exceptions include cold-hardy individuals such as the feijoa, kiwifruit and persimmon. Preference should be given to frost free sites only. Valley lowland and high altitude areas are usually subject to the most severe frosts. Cold night air will always move downslope like an invisible stream and settle in the lowest areas of the garden. These are the danger spots where most cold damage usually occurs. The preferred garden growing sites are elevated in sheltered positions with warm, sunny, northerly or north-easterly aspects. Large bodies of water capture and store heat and have a modifying effect on the local environment. This explains why beachside areas are usually frost free and generally make good growing sites, so long as trees are afforded some protection from the sea breezes and salt spray. If your garden is occasionally visited by frosts, don't worry unduly. There are a number of things you can do to make your growing site warmer and provide a safer environment for your plants to grow in. (Refer to Modifying the Home Garden Microclimate.)

4. PLENTY OF WARMTH AND SUNSHINE

The sun provides light which is essential for the assimilation of nutrients and to improve the colour and flavour of fruit. Sunshine also means warmth, which exotic fruit trees love in large doses. A northerly aspect is ideal as it provides unrestricted sunlight. Try and avoid planting out trees where other larger trees or buildings will compete successfully for their sunlight.

5. ROOM TO GROW

Make sure that the fruit tree you select has enough room to grow and bear fruit, without having to prune it back too much from other trees, buildings, or boundary fences. Excessive pruning may reduce fruit production or lose it altogether for a year or two as the tree becomes vegetative. Before you plant, try to visualise the fully grown tree, how much shade it will cast and its effect on other nearby plants and buildings.

6. ORNAMENTAL BEAUTY

With a little imagination and some careful planning, our home gardens can become more productive, and at the same time they can continue to be a source of much beauty too. For many years the traditional spot to plant fruit trees was in the backyard, right out of view. This was probably the best course of action since our choice of trees was limited mainly to citrus, pomefruit and stonefruits. Times have changed and we are now able to select from a very wide range of trees that not only bear delicious fruit, but make fine ornamentals as well. They all have their own special beauty; whether it be the flash of green and gold from the foliage of caimito, the very showy, crimson-pink 'shaving-brush' flowers of Malay apple, the bell-shaped nocturnal flowers of pitaya, or the golden yellow fruits of carambola. Many of these trees deserve to be in the front garden with your other ornamentals, so why not give them a go?

THE TROPICAL GARDEN

1. Caimito (*Chrysophyllum cainito*)
2. Matisia (*Matisia cordata*)
3. Madrono (*Rheedia madruno*)
4. Mangosteen (*Garcinia mangostana*)
5. Duku/Langsat (*Aglaia domesticum*)
6. Mamey sapote (*Pouteria sapota*)
7. Rollinia (*Rollinia deliciosa*)
8. Mango (*Mangifera indica*)
9. Breadfruit (*Artocarpus altilis*)
10. Rambutan (*Nephelium lappaceum*)
11. Mammea (*Mammea americana*)
12. Pummelo (*Citrus maxima*)
13. Abiu (*Pouteria caimito*)
14. Lime (*Citrus aurantifolia*)
15. Mandarin (*Citrus nobilis*)
16. Carambola (*Averrhoa carambola*)
17. Avocado (*Persea americana*)
18. Jaboticaba (*Myrciaria cauliflora*)
19. Miracle fruit (*Synsepalum dulcificum*)
20. Granadilla (*Passiflora quadrangularis*)
21. Papaya (*Carica papaya*)
22. Purple passionfruit (*Passiflora edulis*)
23. Pejibaye (*Bactris gasipaes*)
24. Salak (*Salacca edulis*)

THE SUBTROPICAL OR WARM TEMPERATE GARDEN

1. Caimito (*Chrysophyllum cainito*)
2. Lychee (*Litchi chinensis*)
3. Grumichama (*Eugenia brasiliensis*)
4. Casimiroa (*Casimiroa edulis*)
5. Acerola (*Malpighia glabra*)
6. Persimmon (*Diospyros kaki*)
7. Macadamia (*Macadamia tetraphylla*)
8. Mango (*Mangifera indica*)
9. Sapodilla (*Manilkara zapota*)
10. Black persimmon (*Diospyros digyna*)
11. Jakfruit (*Artocarpus heterophyllus*)
12. Longan (*Euphoria longan*)
13. Cherimoya (*Annona cherimola*)
14. Guava (*Psidium guajava*)
15. Feijoa (*Feijoa sellowiana*)
16. Carambola (*Averrhoa carambola*)
17. Avocado (*Persea americana*)
18. Jaboticaba (*Myrciaria cauliflora*)
19. Naranjilla (*Solanum quitoense*)
20. Kiwifruit (*Actinidia deliciosa*)
21. Papaya (*Carica papaya*)
22. Purple passionfruit (*Passiflora edulis*)
23. Jelly Palm (*Butia capitata*)
24. Banana (*Musa spp.*)

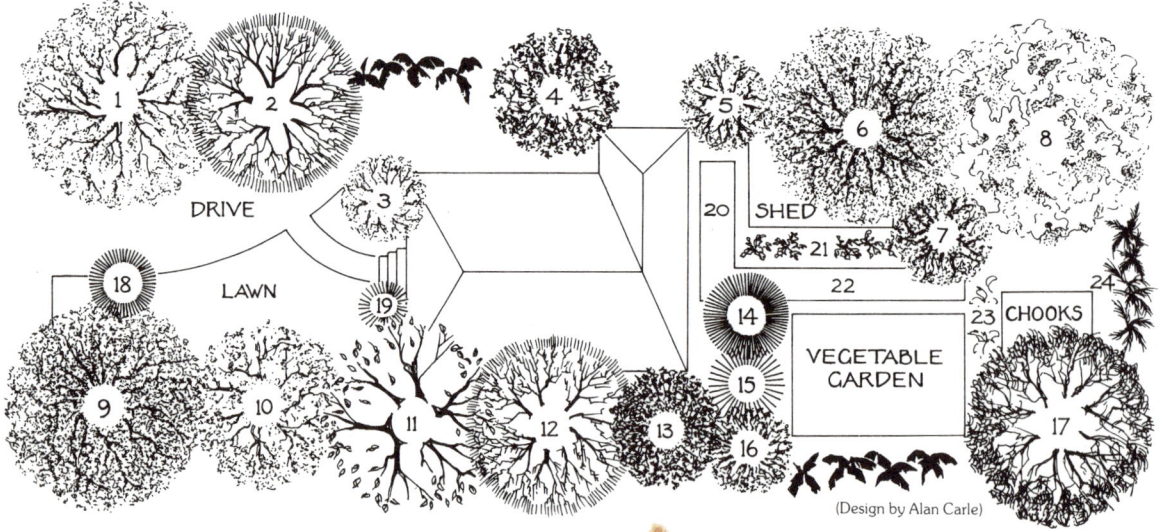

(Design by Alan Carle)

MODIFYING THE HOME GARDEN MICROCLIMATE

With the exception of a few ultra-sensitive plants, most tropical fruit trees will grow successfully in marginal areas outside the tropics. For those of us living in a cooler subtropical or temperate climate, it will probably be necessary for us to modify our home garden environment in order to make growing conditions a little less hostile. Some trees are quite hardy and should grow satisfactorily without too much trouble, however others are a little more sensitive and will need some extra care and attention. There are a number of things we can do to improve the home growing site, and so increase our chances of growing these wonderful fruits successfully.

1. HEALTHY SOIL

A healthy soil should help to get your plants off to a really good start so that they grow vigorously.

Good soil drainage is important, particularly if you are thinking of growing those beautiful but temperamental avocados. They require exceptional drainage, as they are very susceptible to a root-rot disease which thrives in waterlogged soils. Make sure that rainwater run-off from higher ground doesn't enter the growing site by digging interception drains to divert it to other areas. These drains may be open-surface or sub-surface trenches. Sub-surface drains are useful in low lying areas. These are constructed by digging trenches 1 to 1.5 metres deep. Deposit 6 centimetres of fine sand in the bottom of each trench and lay agricultural pipes. Cover the pipes with 12 centimetres of blue metal then add a further 6 centimetres of coarse sand. Refill the remainder with topsoil.

In lower-lying or flat areas where soil drainage is suspect, build up planting sites above surrounding ground levels with good quality topsoil. Garden retaining walls using railway sleepers, stone, brick, or concrete are useful. Make sure you leave adequate drainage or weep holes though, so that there is no water build-up during heavy rains.

Avoid digging holes below ground levels in clay soils and refilling with lighter soils as this commonly forms watersumps which may drown plants. It is possible to improve the structure of heavy clay soils by adding gypsum. Gypsum provides the necessary calcium ions which greatly improve water infiltration rates in clay soils. The calcium pushes sodium off the clay particles and allows them to aggregate together into smaller clumps, around which air and water can circulate more easily. (Apply 0.5 kilogram per square metre.)

Some very sandy soils are too well drained and require extensive watering throughout the growing months to prevent plants suffering from water-stress. Improve the water holding capacity of these soils by adding large amounts of organic matter and some more 'clayey' soil to the planting site. It is a good idea to prepare the soil thoroughly by breaking up all soil clods to a depth of at least 0.5 metres.

Soil fertility in your garden will be greatly increased by adding organic matter. Dig in large quantities of compost and well rotted animal manure to the soil in each planting site. This will build up organic nitrogen levels and also stimulate production of high levels of microbial activity. Large numbers of soil micro-organisms will compete with, and actively antagonise, the root rot fungus associated with *Phytophthora cinnamomi*, should it be present. They will also aid in the breakdown and release of nutrients from the organic matter. Earthworms are to be treasured. They maintain good soil structure by improving drainage and aerating the soil. They drag down nutrients into the root zone, and increase soil nitrogen levels by deposition of skin excretions (slime), castings, and used carcasses. Most exotic fruit trees grow best in soils with a pH of 5.5-6.5. In many cases soils are too acidic and require lime or dolomite to raise the pH to more comfortable levels.

2. SHELTER FROM DAMAGING WINDS

Exotic fruit trees of all ages are damaged by strong winds. Dessication of young, tender growth, broken limbs and bruised fruit commonly result. In many suburban gardens windbreaks are already present in the form of buildings and boundary fences. These are not very effective as windbreaks, as they often cause harmful wind tunnel and eddying effects. We should aim at providing natural shelter by growing hedges of shrubs and small trees which filter the wind. Wind shelter plants may be productive or non-productive. If home orchard space is limited, then a productive windbreak formed from various types of fruiting plants is the obvious choice. Plants are grown along boundaries as a first line of wind defence, and also within the orchard to provide increased protection. For boundary windbreaks, native species of melaleuca, callistemon, leptospermum, and casuarina provide some quick cover. In warm areas the jakfruit and jambolan make effective windbreaks when close-planted in hedgerows, and provide some tasty fruit as well. Further wind protection can be provided by growing plants such as bananas, corn, or sorghum amongst the fruit trees within the home orchard. This also helps to create a warm, steamy, tropical microclimate ideal for fruit production. Artificial windbreaks and shelters are also useful. If you don't have the room to plant hedgerows, then one alternative is to obtain some windbreak or windfilter cloth and erect it around the orchard perimeters. In more exposed gardens, young plants can be given extra protection for the first twelve to twenty-four months by constructing individual climate shelters with the use of garden stakes, hessian or synthetic materials such as plastic, shadecloth, and fertiliser bags.

3. COLD PROTECTION

Young trees less than two or three years old are most susceptible to low temperature damage. As trees mature and grow larger they are better able to withstand these cold conditions. Frost protection in the early years can mean the difference between success or failure.

The coldest nights are those when the skies are clear and there is no cloud cover. Under these conditions, maximum heat is radiated from earth back into space. The main danger period for frosts is in the early hours of the morning just before sunrise.

One of the most effective frost protection methods is to cover the tree at night and trap the heat that re-radiates from the soil beneath it. For small plants a paper bag placed over the top can help. Individual climate shelters using plastic coverings are very effective (make sure that air circulation is provided during the daytime, or your plants may 'cook'). Suspending shadecloth or mosquito netting over trees and hanging newspaper or banana leaves in amongst the branches also helps to retain heat. Tree trunks just above ground level are cold sensitive and can be wrapped in cardboard for added protection. Surface mulches under the tree canopy should be raked clear as their insulative properties tend to prevent heat from leaving the soil and rising up into the tree. Moist soils are better heat conductors than dry soils, however watering soils deliberately should only be carried out if the ground is dry and when warm weather is predicted for the following day. Cold, wet soils can damage trees more than the frosts themselves. Leak-proof containers are effective when filled with water and placed under the tree canopy. The water releases heat and occupies the lower space that is normally occupied by cold air. Household sprinklers can be turned on in the early hours of the morning and left on until the frost danger has passed. Wind generators and fans mix warm and cold air together and this can raise the temperature by a valuable one or two degrees. Smudge pots that burn oil-soaked sawdust can create a cloud of warm air that drifts through the orchard raising temperatures. Make sure that any windbreaks on the low side of the orchard have adequate gaps in them to allow the cold air to continue in its downslope direction. If you have a small greenhouse, it may be a good idea to keep your more sensitive fruit trees there until they are 1 to 2 metres tall, and then you can plant them out in the orchard. At that stage they are bigger and stronger and should be better able to cope with cold spells.

TROPICAL
FRUIT

ABIU

(*Pouteria caimito*)

From the headwaters of the Amazon and its tributaries in tropical Brazil and Peru comes the abiu, *Pouteria caimito*. This scrumptious fruit is a real taste-treat when eaten slightly chilled. Just slice the fruit in half and spoon out the sweet, succulent, creamy-white flesh. Abius are round to oval in shape, sometimes pointed, and have very attractive, smooth, bright yellow skins. They are borne directly onto the stems of the larger twigs and tree branches.

The abiu is a small evergreen fruit tree with smooth, mid-green leaves clustered towards the ends of the twigs and willowy branches. Trees grow to a height of up to 16 metres in the uniformly warm and humid climate of the Amazon Valley. However, in Australia they are more commonly seen as small orchard specimens reaching some 4 to 8 metres. They often crop several times during the year. Other common names used for the abiu include **caimo** and **cauje**.

MICROCLIMATE

Abius prefer a warm, humid climate and may be grown satisfactorily in most tropical and warmer subtropical areas. The young plants are very cold tender, usually not tolerating frost conditions when mortality rates may be high. As trees grow and become established they are more cold hardy and usually survive. Wind protection for young plants is beneficial. Trees adapt to a wide range of soils, providing it is well drained and fertile. They also appear to have some drought tolerance.

VARIETIES

There are few named varieties at present. Seedling trees may take two to three years to bear, however some fruit in eighteen months, or less, from planting out.

CULTURE

Plant out in a sunny spot at least 4 metres from existing trees or buildings. Plantings should be carried out in the warmer months of the year, preferably during periods of expected good rainfall. Water in well to settle the soil around the roots. No fertiliser should be given at planting time. Mulch well around all trees and apply frequent light applications of poultry manure and blood and bone. Be careful not to place mulches too close to the tree trunks. Water frequently during warmer months, particularly from fruit set onwards. Young abius should be given frost and cold wind protection. Construction of individual climate shelters may be required in more exposed locations. Only light pruning is required. This is chiefly carried out when plants are still young, to provide the tree with a good framework of well spaced branches instead of a central leader.

HARVEST

Abius should be harvested when fully mature, as immature fruits contain an unpleasant, milky latex near the skin which sticks to the lips after eating. Allow them to ripen fully before consuming. Where nylon bagging of fruit is used (fruit fly susceptible areas), it is a good idea to wait until the fruit drops naturally into the bag before harvesting. The main harvest season occurs between January and September with some trees having up to three crops during this period. Fruit should not be frozen, but will store satisfactorily in the refrigerator for a few days.

PROBLEMS

No serious diseases affect production. The fruit fly may be an occasional pest.

CULINARY USE

Abius are very tasty eaten fresh by themselves. In fruit salads, their flavour is enhanced by the addition of a slice of orange to provide a little acidity, which the abiu lacks.

ACEROLA

(*Malpighia glabra*)

Acerola's bright red, thin-skinned, cherry-like fruits are an exceptionally rich source of vitamin C, containing between twenty and fifty times as much as the orange! Taste varies from acid to sweet depending upon variety. The highest vitamin count is present in the sour, unripe fruits. Fruits become sweeter as they ripen and are eaten raw, used in drinks, or made into tasty jams.

The acerola, *Malpighia glabra*, is also commonly referred to as the **Barbados cherry** and comes from the West Indies and Central America. It is a large, densely branched shrub or bush which grows to a height of 3 to 4 metres. A very decorative specimen with its dense cover of glossy, dark green leaves, the acerola is appreciated almost as much for its ornamental value as it is for its fruit.

MICROCLIMATE

Acerola may be grown successfully throughout most tropical and subtropical areas. Young shrubs are frost tender, however when mature they are more cold hardy. Plants grow on a wide range of soils, but prefer the richer, free-draining types containing plenty of organic matter. A sunny, sheltered position is best.

VARIETIES

Seedlings are not recommended due to their variability. Clonal varieties have been selected for their high yields, and their sweet and juicy fruit. Some varieties include: CALIFORNIA HONEY, FLORIDA SWEET and B17.

CULTURE

Plant out in warmer months (preferably during periods of expected good rainfall), at least 2 metres from established trees or buildings. Mulch well to help conserve soil moisture, retard weed growth, and encourage soil micro-organisms. Individual climate shelters in more exposed gardens are a good idea to protect young plants from the dangers of frosts and cold winds. Fertilise in early spring and through summer using well rotted animal manure mixed with mulch. Water well during dry spring periods, when bushes are flowering, and maintain a good supply right through to the end of the harvest season. Prune in early autumn, thinning dense growth at the top of the bush to promote side branching.

HARVEST

Plants should begin to bear in their second year from planting out. The harvest season may extend over several months from spring through to autumn. Fruits mature three or four weeks after each flowering flush. They are ready to pick when they turn deep red. If picked half ripe they will store satisfactorily for several days in the refrigerator. Frozen juice keeps well and retains its flavour, colour and vitamin content. Fresh fruit life is short and fruits should be used as soon as possible after harvest.

PROBLEMS

Pests include scale insects, plant bugs, butterfly larvae, and nematodes. No major diseases affect production.

CULINARY USE

Acerola fruit is commonly eaten raw as a table fruit. It is also used in punches, preserves, jellies and purees.

1	Mango	**24**	Imbe
2	Watermelon	**25**	Breadfruit
3	Honey dew melon	**26**	Pineapple
4	Rockmelon	**27**	Lychee
5	Santol	**28**	Jakfruit
6	Mangosteen	**29**	Guanabana (soursop)
7	Cocona	**30**	Uvilla
8	Rambutan	**31**	Sweetsop
9	Jelly palm	**32**	Sapodilla
10	Yellow passionfruit	**33**	Ambarella
11	Purple passionfruit	**34**	Persimmon
12	Abiu	**35**	Cashew
13	Casimiroa (white sapote)	**36**	Kitembilla
14	Bilimbi	**37**	Carambola
15	Acerola	**38**	Tamarillo
16	Mabolo	**39**	Longan
17	Cherimoya	**40**	Banana
18	Bell fruit	**41**	Papaya (Papaw)
19	Macadamia	**42**	Pitaya
20	Avocado	**43**	Monstera
21	Prickly pear	**44**	Pummelo
22	Durian	**45**	Atemoya (Custard apple)
23	Black persimmon	**46**	Marang

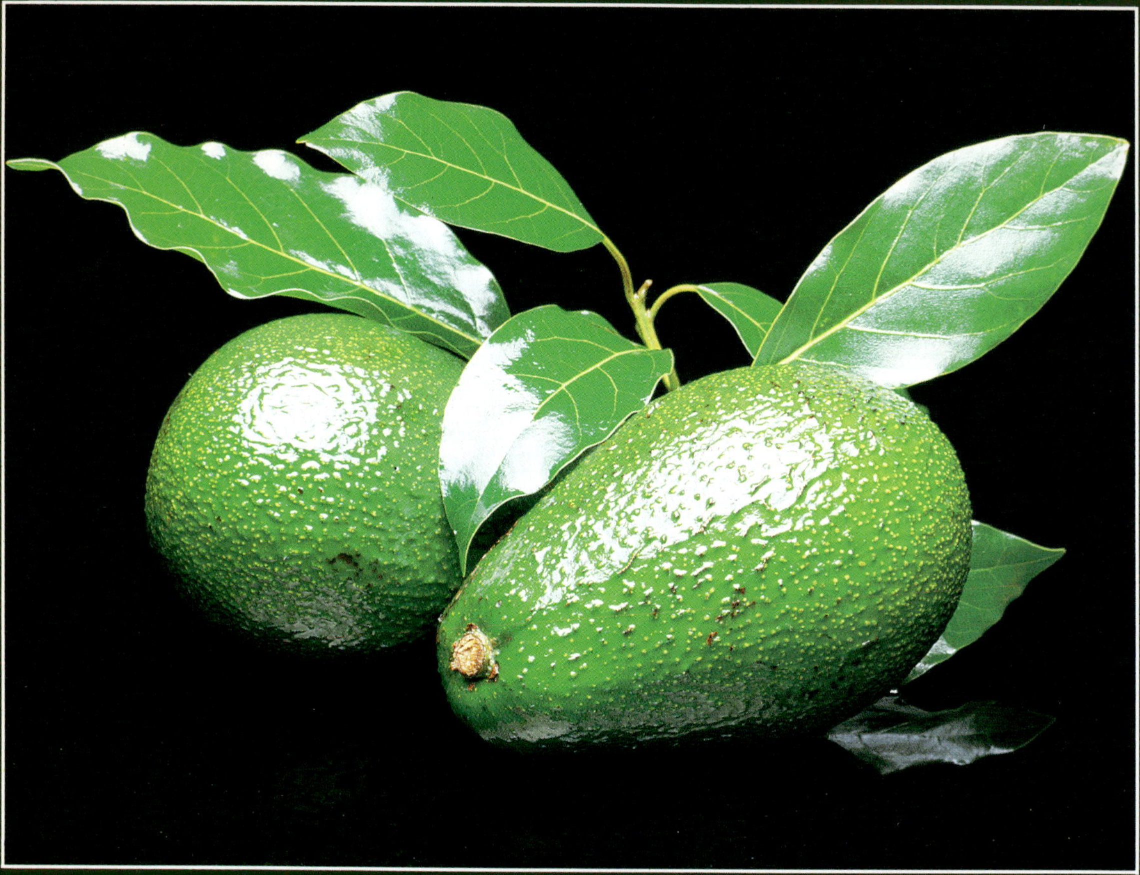

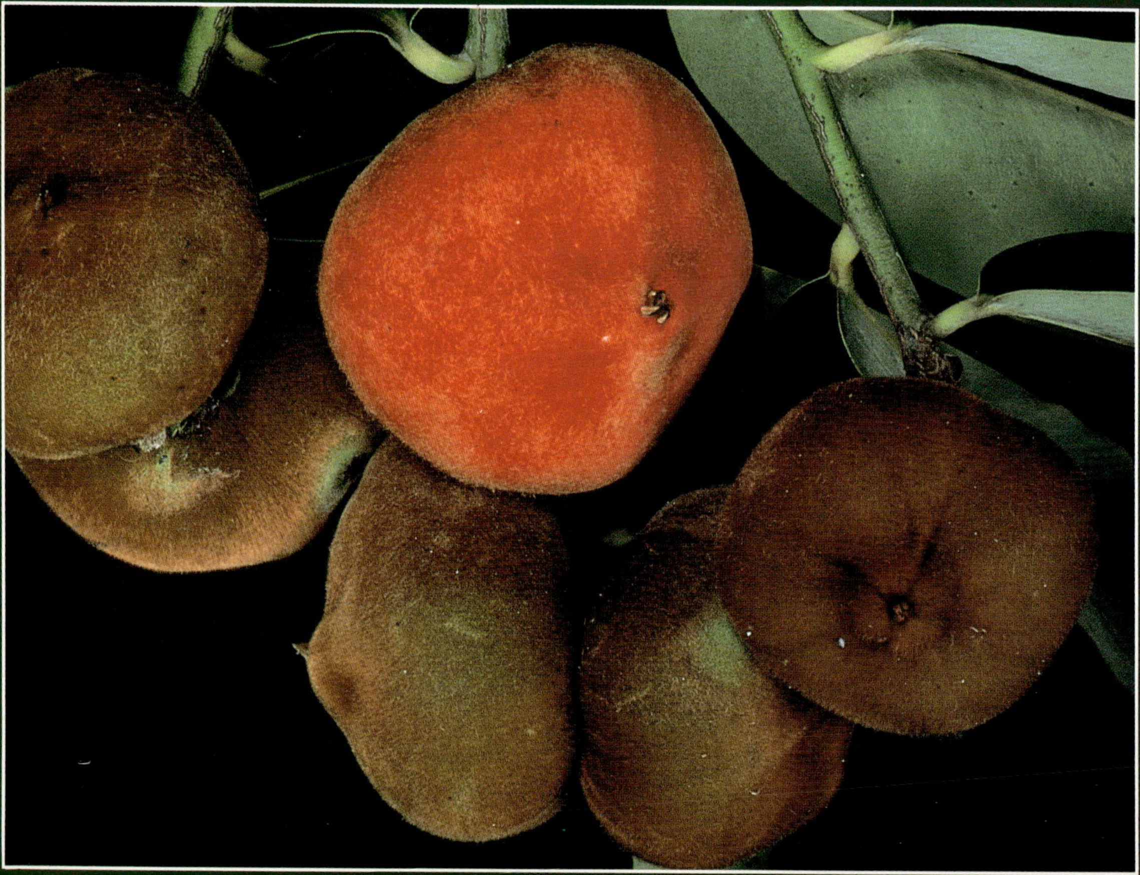

Top left
Avocado. (T. Hoult/MPA)

Left
Mabolo. (J. Truscott)

Above
Rollinia. (D. Cilento/QTTC)

Papaya, mangosteen, mango, rambutan, duku-langsat, salak, guava. (R. Francis)

ATEMOYA
(*Annona atemoya*)

Custard apples, *genus Annona*, originated in the tropics or near tropics of Central and South America and there is evidence of their cultivation in prehistoric Peru. They are now cultivated commercially in many tropical and subtropical areas throughout the world. There are several species in the custard apple family including the **cherimoya**, *A. cherimola*, the **sweetsop**, *A. squamosa*, the **atemoya**, *A. atemoya*, the **guanabana**, *A. muricata*, the **ilama**, *A. diversifolia*, the **mamon**, *A. reticulata*, and others.

The atemoya is a natural hybrid of the sweetsop and the cherimoya, and to which the common name custard apple often applies. It grows to be a small, spreading tree or shrub to a height of 5 to 7 metres. Trees are semi-deciduous, losing leaves in winter-spring prior to new growth. The fruits vary from heart-shaped to rounded varieties and weigh up to 2 kilograms each. They have a light green skin and a white, juicy flesh with an agreeable blend of mild acidity and sweetness. Numerous brown or black seeds are distributed throughout the flesh.

MICROCLIMATE

Atemoyas prefer a warm to hot climate, with relatively high rainfall and humidity during flowering. Trees may be grown on a wide range of soil types, providing they are fertile and well drained to discourage root rot diseases. Shelter from strong winds is desirable as the large, broad leaves tend to act as sails, often resulting in limb breakage. Mature trees withstand light frosts, however the fruit is usually damaged. Young trees may not survive unless given some frost protection, such as wrapping trunks in cardboard.

VARIETIES

The main varieties grown in Australia are AFRICAN PRIDE and PINK'S MAMMOTH. African Pride produces heart-shaped, symmetrical fruit compared with the more irregularly shaped Pink's Mammoth. It usually comes into bearing at an earlier age, with higher yields and less vigorous tree growth. However Pink's Mammoth produces the sweeter and more aromatic fruit. Other varieties include: NEILSEN, QAS, WHITE, BULLOCKS HEART and GEFNER.

CULTURE

Plant out in early spring (after the last of the frosts), at a distance of not less than 4 or 5 metres from existing trees or buildings. Mulch well to help conserve soil moisture. Trees may need staking in more exposed gardens. No artificial fertiliser should be given in the first year. From the second year onwards apply 0.5 kilograms of complete fertiliser for every year of growth in split applications in September, December, February, and April of each year. Long, whippy branches should be pruned back about 1 metre each year, and dense crowns should be thinned to enhance internal bud formation. Water well from bud break in spring right through to harvest in autumn. Be careful not to overwater as atemoyas are susceptible to root rots.

HARVEST

Grafted trees should commence bearing when between two and four years old. Atemoyas are harvested in March in northern areas, with the season continuing until August-September in southern areas. A yellowing of the space between some of the corrugations on the fruit surface heralds harvest time. The fruits should be ready to eat about four to seven days after picking. The best time to eat them is about two days after they have softened. Do not store them at less than 15°C before they soften, as cold storage damage may occur (a temperature of about 17°C is thought to be optimum).

PROBLEMS

Main pests include fruit fly, mealy bugs, scale insects, leaf eating caterpillars and the yellow peach moth. Severe collar and root rot problems can result from the action of a bacterial wilt, *Pseudomonas solacearum*.

CULINARY USE

The creamy-white, sweet flesh is eaten on its own or mixed with other salad fruits. It also makes tasty ice-cream, sherbets, fruit drinks and pies.

AVOCADO

(*Persea americana*)

This versatile and highly nutritious fruit is native to the highlands of Mexico and Central America and the lowlands of Colombia and South America. When the Spanish invaded the Incan and Aztec territories in the sixteenth century they found it growing extensively and it was noted to play an important role in the diet of the natives. Avocado fruit has a high content of mono-unsaturated oil (5-30%) and contains no cholesterol. Apart from the obvious nutritional benefits, there have also been unconfirmed reports of this fruit being used successfully as an aphrodisiac among the pygmy tribes of the lower Amazon Basin. It is said that they even prefer it to marijuana, which is also readily available!

Avocado flesh is soft or deep green, not very sweet, with a delicate flavour, and when ripe possesses a nutty, slightly herby flavour and dissolves like butter on the tongue. Under favourable conditions the avocado, *Persea americana*, grows to be a large evergreen, with mature specimens reaching 10 to 12 metres or more and having a diameter of up to 13 metres. There are three main types of avocado; **Mexican**, **West Indian**, and **Guatemalan**. Most varieties grown in Australia are of Mexican or Guatemalan origin, or hybrids of these.

MICROCLIMATE

Avocados require an absolute minimum of 1.5 metres of well drained topsoil, above a porous subsoil, as trees are very susceptible to *Phytophthora cinnamomi*, a root rot fungus which thrives in waterlogged soils. An elevated growing position away from frosts is recommended, although when mature trees usually tolerate light frosts. A temperature range of 20-25°C is optimum for root growth, flowering, and fruit set. A warm, sunny location sheltered from strong winds is ideal.

VARIETIES

Due to their unusual flowering habit, avocado varieties may be divided into two classes, A and B. In most home garden situations there is usually sufficient cross pollination on one single tree to produce enough fruit. If you are having fruiting problems you may need to interplant varieties from Class A and B whose flowering times overlap. The presence of bees also aids pollination. Some Class A varieties include: HASS, REED, WURTZ, HAZZARD, RINCON, MACARTHUR. Some Class B varieties include: FUERTE, SHARWIL, NABAL, EDRANOL, BACON.

CULTURE

Plant out in the warmer months of the year at least 5 metres from existing trees or buildings. Mulch well to discourage weed growth and encourage large amounts of soil micro-organisms. It is a good idea to protect young plants with individual climate shelters. Provide some shade over the top for the first few summers. Prune tall, upright varieties such as Hass by removing the apical bud of the leaders to give trees a more compact, spreading habit. Spreading varieties such as Fuerte are pruned by trimming back lateral branches to encourage upward growth. A tablespoon of complete fertiliser should be given to young growing trees every eight weeks. Poultry manure and gypsum are very useful when added to the mulch. Aliette® will help prevent root rot disease. Water trees at least twice a week during the growing season, but only moderately during winter.

HARVEST

The avocado is rather unusual in that it matures on the tree but does not ripen until after it is picked. Therefore harvesting time is not critical. Hass varieties change colour from green to purple, and the fruit stems of Fuerte varieties yellow slightly. In all varieties a shiny appearance may be replaced with a dull green skin colour. Picked fruit should ripen within ten days at room temperature without wilting. To retard fruit ripening, green mature fruit (just picked from the tree) may be stored at 7°C for several days without serious chilling injury. Do not store at lower temperatures. Fruit should be placed under refrigeration not more than twenty-four hours after harvest. Ripe fruit may be held at 0°C for up to seven days.

PROBLEMS

The main problem is caused by a soil fungus, *Phytophthora cinnamomi*, which results in root rot. Other diseases include anthracnose, sunblotch, verticillium wilt, stem canker and bacterial soft rot. Pests include fruit fly, monolepta beetle and fruit spotting bug.

CULINARY USE

Avocados are popular eaten alone as a dessert dish. They are also used to make tasty savoury fillings, soup, ice-cream, butter, garnishes and salad dressings. Their flavour is enhanced by sprinkling them with lemon or lime juice, sugar or spices, or simply seasoning with salt and pepper.

BABACO

(*Carica pentagona*)

The babaco, *Carica pentagona*, belongs to the same family as the papaya and comes from the highland regions of the Andes mountains in Ecuador. It is a small, subtropical, herbaceous tree or shrub, growing rapidly to a height of 2 to 2.5 metres. The leaves are produced in similar fashion to the papaya, extended on long stalks and forming a crown or canopy near the top of the quick-growing, fleshy stems. Prolific quantities of bright yellow skinned fruits are produced for three to six months of the year.

Babaco has a highly aromatic, juicy, rather acid, cream coloured flesh, free from seeds. The skin as well as the flesh may be eaten. The taste is rather fruity, reminiscent of a blend of pineapple, papaya, and strawberry. In Ecuador, juice of the babaco makes a popular drink and is served in the top hotels. Plants are often cultivated intensively on smallholdings, along with other activities such as free-range chicken farming.

MICROCLIMATE

The babaco is more cold tolerant than the papaya, and grows well in its native Ecuador where the average maximum temperature is 20°C and minimum is 5°C. A cool subtropical climate free from frosts is ideal. Plants are sensitive to high temperatures and low humidity which may result in sunburnt fruit and immature fruit drop. The babaco is susceptible to root rot so soils should be friable and free-draining to a depth of 0.5 to 1 metre, and contain good quantities of organic matter. Wind protection is essential as plants have a shallow root system and may blow over when laden with fruit.

VARIETIES

The babaco must be propagated vegetatively by stem cuttings and tissue culture, as the fruit is seedless (no male trees or pollen are required for babaco to set fruit). There are no separate varieties.

CULTURE

Plant out in spring months at least 1 metre from existing trees or buildings. Full sunshine is best for fast, productive growth. Mulch well and apply frequent light dressings of poultry manure spread evenly under the leaf canopy, but not too close to the stem. Water well in spring and summer months, however be careful not to overwater as root rot may result. Prune the stem back to 15 to 25 centimetres above ground level after harvest each year, and allow one or two shoots to grow up as next season's fruiting stems.

HARVEST

The first fruit ripens when plants are between nine and fourteen months old. Fruit that is set below 30 centimetres from ground level is normally removed as it may rub on the ground. A single babaco tree can produce twenty-five to one hundred fruits each year. Fruit harvested when about 30% yellow has a shelf life of up to four weeks, and even more if kept in the refrigerator. Babaco is best eaten when fully ripe.

PROBLEMS

Pests include spider mites, two-spotted mites, broadmites, and nematodes. Diseases include powdery mildew and *Phytophthora* root rot.

CULINARY USE

Babaco may be eaten on its own or mixed with other fruit. Juice can literally be poured from the fruit and makes a refreshing sorbet with lemonade. The fruit can also be baked, bottled, stewed, barbecued, or made into an excellent preserve.

BANANA

(*Musa spp.*)

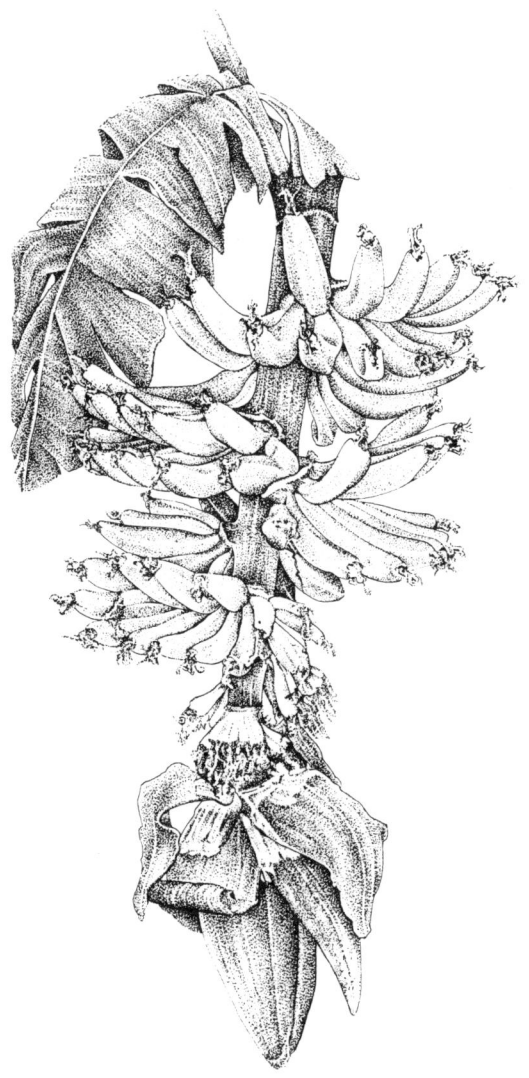

Bananas not only taste terrific, they're also highly nutritious, and for this reason they are one of the most important economic food producing plants in the world. They contain plenty of protein, vitamins A, B and C, and minerals, including iron, magnesium, phosphorus and potassium.

The banana, *Musa spp.*, is a large, tree-like, herbaceous plant native to South-East Asia, where some wild forms *M. acuminata* and *M. balbisiana* are thought to have given rise to the commercial varieties grown today. The most common type of banana is the dessert banana, eaten fresh for its fine, sweet flavour. Another type is the plantain which is usually cooked first as the raw flavour is often unpleasant and rather tasteless.

Wild plants up to 6 metres tall may still be found growing throughout the tropical jungles of South-east Asia. Their fruits are mostly small and hard, and contain numerous black seeds.

MICROCLIMATE

Bananas grow best in a hot-wet tropical climate with mean monthly temperatures of 27°C. Frost is not normally tolerated at any stage, and plants may suffer cold damage when temperatures fall below 5°C. A warm, frost free site facing north and sheltered from cold southerly and westerly winds is recommended for good growth and fruit production. The banana can grow on a wide range of soils as long as they are reasonably fertile and not subject to constant waterlogging. The best soils are volcanic or alluvial in nature, deep and well drained, with a high humus content and a pH 5.5-6.5. Plants require high soil moisture, preferably between 20 and 50 millimetres per week. A growing position in full sun is recommended.

VARIETIES

In warmer areas flowering usually occurs six to eight months after planting and fruit matures some three to five months later. Bunches are produced every nine to twelve months on new stems. Some varieties include: WILLIAMS, CAVENDISH, LADY FINGER, SUGAR, GROS MICHEL, PISANG RAJA, PISANG MAS and DUCAS.

CULTURE

Plant out in spring months at least 1 metre from existing trees or buildings. Suckers should be planted upright in holes dug about 40 centimetres deep. Unless conditions are very dry don't water for two or three weeks as the planting material may rot. Mulch well to reduce competition from weeds. No fertiliser should be given for the first two to three months, however from then on a complete 'banana fertiliser' is recommended. Allow only one sucker to grow to replace the mother or fruiting stem. This should have a thick base, tapered stem and thin leaves. Plastic covers are used about three weeks after a bunch has emerged. These are tied loosely at the top to allow some air circulation, and left to hang open at the bottom. They are used to help fill bunches faster and to protect fruit against blemish from leaf rub and pests such as the fruit fly. Cut off the bell about 10 centimetres below the last hand to increase fruit size. As the fruiting stem only produces one bunch in its life cycle, it should be cut down to ground level after harvest.

HARVEST

Harvest bunches when the top hand starts to change colour and the fruit is full. Hang them in a cool ventilated spot out of the sun. They may ripen all at once, so it is a good idea to break off one or two hands at a time, from the top of the bunch. These should ripen ahead of the main bunch.

PROBLEMS

Pests include fruit fly, nematodes, banana weevil borer, rust thrips, flower thrips, red spider and fruit eating caterpillar. Diseases include leaf spotting, leaf speckle, panama disease, and bunchy top.

CULINARY USE

Bananas can be used in fruit juices, salads, desserts, pies and all baking uses. Peeled bananas will soon turn brown. This process can be slowed down by sprinkling with lemon juice.

BLACK PERSIMMON

(*Diospyros digyna*)

Mexico is home to the black persimmon, *Diospyros digyna*, where it grows to be a very handsome evergreen fruit tree with shiny, dark green foliage and round, olive-green skinned fruit. The dark chocolate-brown flesh within has a very rich, sweet flavour that is enhanced by the addition of a little vanilla, rum or lemon juice. It makes delicious ice-cream and is also very tasty in mousses, cakes, bread and preserves. Harvested fruit remain quite firm for several days then ripen marshmallow-soft, almost miraculously, overnight.

Trees are hardy and grow with plenty of vigour to attain a height and spread of 6 to 9 metres at maturity. Black persimmon fruit is an excellent source of vitamin C, containing about four times as much as the orange. It also has good amounts of calcium and phosphorus. There are seeded and seedless types.

The black persimmon is cultivated as a minor crop in Florida, Hawaii, Mexico, California and the West Indies. Other names for the fruit include **black sapote**, **chocolate pudding fruit**, **chocolate persimmon**, **guayabote**, and **sapote negro**. Close relatives of the black persimmon are the mabolo, *D. discolor*, and the persimmon, *D. kaki*.

MICROCLIMATE

Trees prefer sun or semi-shade, and will grow well in a wide range of soils providing they are moist and well drained. The black persimmon is adaptable to tropical and subtropical climates, however young trees may be killed at temperatures below 0°C, and mature trees killed or severely damaged by temperatures of −2°C or less for prolonged periods. Protection from strong winds is important as tree branches are fairly brittle.

VARIETIES

Some seedling trees begin to bear about three years from planting out. There have been limited selections made overseas, with very few grafted trees being available here at present. REINEKE is a popular variety from Florida.

CULTURE

Plant out in a sunny, sheltered spot at least 4 metres from existing trees or buildings. Water in well to settle the soil around the roots, and mulch well to reduce weed competition and help conserve moisture. Frost and wind protection is advisable for young trees up to two or three years old so that they will grow unchecked. Each year after harvest, trees benefit from an addition of dolomite or lime, a dressing of poultry manure and a light mulch cover. Soil moisture should be maintained in dry periods, however, apart from this trees are best left alone. Too much tender loving care, particularly in more tropical locations, often leads to vegetative growth rather than quality fruit production.

HARVEST

Harvesting is carried out from autumn to early spring. Fruit does not change colour (although there can be a slight darkening of segment lines), and remains firm when mature on the tree. Choosing the time to pick depends mainly on fruit size. Allow fruit to ripen in a cool, dry place. This may take from seven to fourteen days. Keep a close watch as it ripens quite suddenly after remaining hard for days. Cut the fruit open from blossom to stem end into four or six slices or wedges. Gently scoop out the pulp, removing the seeds. Eat fresh or freeze pulp straight away. Green unripe fruit can be held at 10°C for several months and should soften up in a few days at room temperature.

PROBLEMS

No serious pests or diseases affect production. Scale insects may cause some minor problems.

CULINARY USE

Black persimmon mousse is now becoming popular in restaurants. The fruit also adds flavour and texture to ice-cream, preserves and cakes.

BREADFRUIT

(*Artocarpus altilis*)

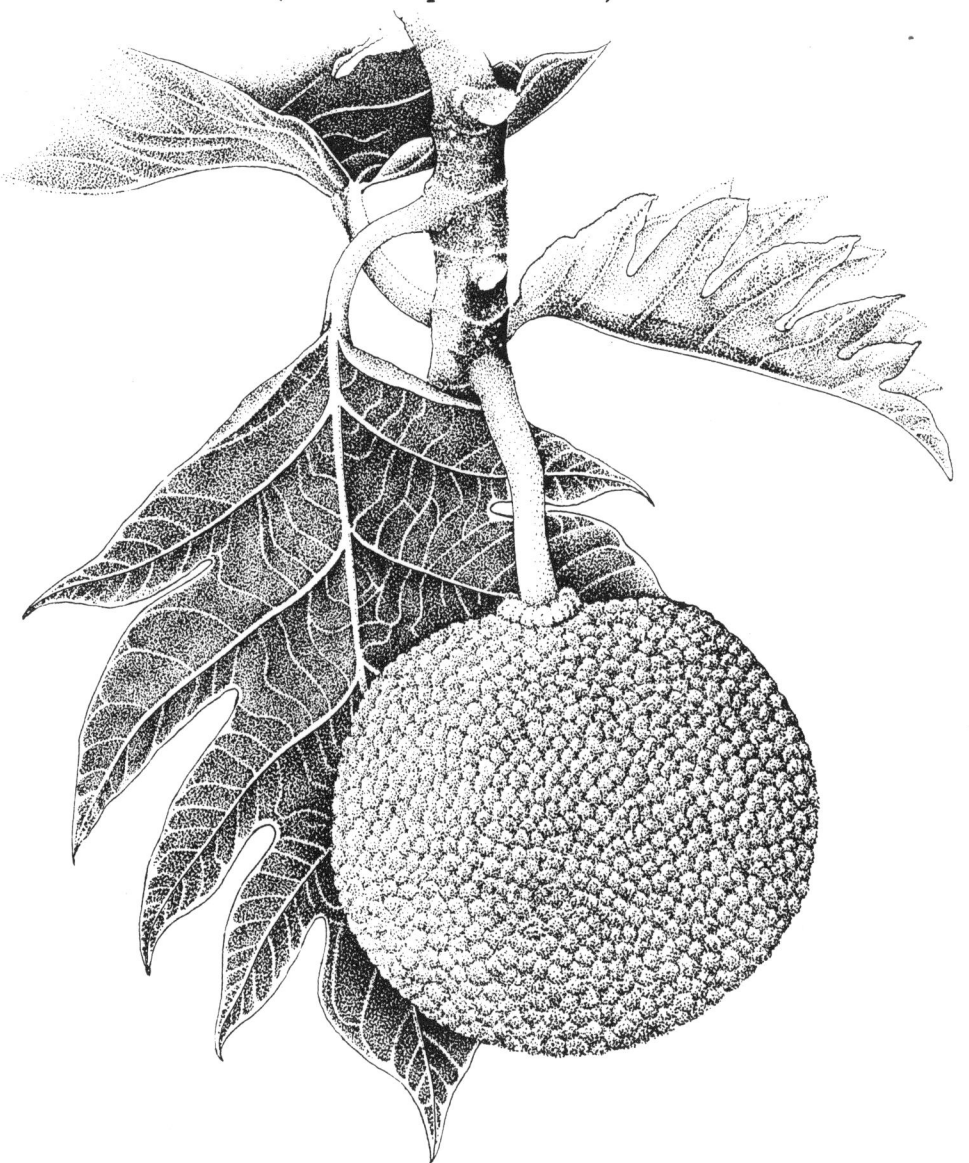

The breadfruit, *Artocarpus altilis*, is believed to have originated in South-East Asia where a wild species has probably given rise to the domesticated species more commonly grown today. Breadfruit and other plants such as cassava, taro, and sweet potatoes are important starch staples for the inhabitants of tropical regions.

Breadfruit trees look very ornamental in the home garden with their large, deeply lobed, dark green leaves which provide some cooling shade in the hot summer months. Mature trees grow to a height of 8 to 20 metres depending on locality.

The round, green fruit has a hard, seedless flesh which is usually cooked as a vegetable. When ripe, the flesh is soft, sweet and yellow. There is also a seeded type known as the breadnut; it has little edible flesh but the seeds are popular roasted, and taste like chestnuts.

MICROCLIMATE

Trees grow best in damp, hot, tropical lowlands. They are a tropical species and are very cold sensitive, not tolerating frost of any description. Some leaf drop may occur during strong winds, dry spells, and when temperatures fall below about 8°C. Young trees benefit from protection from the hot sun until they are established, however after this they prefer full sunshine. Trees have some wind and salt tolerance. Breadfruit will grow on a wide variety of soils providing that they are deep and well drained. Good growth is promoted in high fertility soils containing good amounts of organic matter.

VARIETIES

Breadfruit trees should only take three years to bear fruit if they are given good growing conditions. They continue to bear for thirty to forty years. Some varieties include: LARGE GREEN, SOLOMON YELLOW, SAMOAN and CRICKET BALL.

CULTURE

Plant out in warmer months in a position sheltered from cold winds at least 5 to 7 metres from existing trees or buildings. Mulch around each tree and apply frequent light dressings of poultry manure as fertiliser. Water plants regularly, particularly during long, dry spells. Minimal pruning is necessary other than removal of dead or dying branches.

HARVEST

Breadfruit may be harvested at the immature stage when the milky sap comes to the surface and the fruit is still firm, green and starchy or at a later stage when the rind turns yellow-green with a sweeter taste. Breadfruit is nearly always cooked before it is eaten. Sometimes it is fermented by burying it in layers between the leaves. It is then mixed with coconut cream and baked into bread. The shelf life can be extended up to two weeks by cool storing fruit wrapped in polythene bags at 12°C. It can be preserved for longer periods by drying or burying. Polynesians usually cook breadfruit before storage.

PROBLEMS

Flying foxes, termites, and brown scale may be occasional pests. No diseases seriously affect production.

CULINARY USE

Breadfruit is traditionally baked in ground ovens or roasted over hot coals. It is also rather nice when cooked the same way as potatoes; boiled and mashed with milk and butter. Breadfruit chips are very popular; the breadfruit is sliced thinly and then fried in oil until crisp. Other uses include puddings, bread and for pie making.

CAIMITO

(*Chrysophyllum cainito*)

Apart from its handy habit of bearing delicious fruit, caimito also makes a fine ornamental specimen for the home garden with its graceful shape and beautiful two-tone foliage. Leaves are a dark, glossy green above, with an undersurface of silky golden brown, and look very striking as they rustle gently in the breeze. Caimitos are round with smooth, thick, pale green, purple or copper coloured skins. The flesh is sweet, white and semi-translucent and usually contains between two and five seeds. Slicing the fruit horizontally reveals an attractive star-shaped pattern.

Caimito, *Chrysophyllum cainito*, is native to the West Indies and Central Americas where it is a popular fruit tree among the local inhabitants and is cultivated extensively. It is a medium to large evergreen, growing to a height of up to 15 metres under favourable conditions and requires little or no pruning. The tree is also very popular in Thailand and the Philippines. Other common names for caimito include the **star apple** and **sawo duren**.

MICROCLIMATE

Caimito is a tropical tree, however it grows in warm, subtropical climates too. When mature it is fairly frost hardy and will tolerate light frosts with only minor damage resulting. Young trees are not so cold hardy and must be given frost and cold wind protection for the first two or three years. A rich, well drained soil is preferred, however trees usually grow well on a wide range of soil types. Young plants benefit from shade until they have become established, then they grow best in full sunshine.

VARIETIES

Seedling trees are often irregular bearers, sometimes producing no fruit at all, and the fruit that is produced can be poor quality. Grafted trees are more reliable and are grown in preference to the seedling types. Some varieties include: GRIMAL, HAITIAN, PHILIPPINE GOLD, NEWCOMB, WEEPING and PUBLICO.

CULTURE

Plant out in warmer months in a warm, sheltered position at least 4 or 5 metres from established trees or buildings. Mulch well, but make sure that the base of the trunk is kept clear. Provide individual climate shelters in more exposed gardens. Water well during the growing months and apply frequent light applications of animal manure to the mulch as fertiliser. Little or no pruning is necessary other than to remove any dead or diseased branches, and to prune trees to grow with a single trunk.

HARVEST

Trees should begin to bear fruit in their third or fourth year. Caimito is a prolific cropper, and yields on some young trees can be very good (up to a hundred fruits in one season). Fruit mature from mid-winter through to early summer. They should be allowed to ripen fully on the tree as immature caimitos have an unpleasant sticky latex beneath the skin.

PROBLEMS

No major pests or diseases affect production.

CULINARY USE

Fruit are eaten on their own, or they may be mixed with other fruits such as citrus. Caimito makes a good drink when blended with orange. They are also used in sorbets, creams and jams.

CANISTEL

(*Pouteria campechiana*)

The canistel, *Pouteria campechiana*, was well known to the pre-Columbian civilisations of Central and South America, and was depicted on clay pots and woven cloth. An open-growing, evergreen fruit tree from Mexico and Central America, it reaches a height and spread of 8 to 10 metres with long, smooth, dark green leaves clustered near the ends of the twigs and smaller branches.

The canistel is a prolific bearer of roundish, pointed fruit with a smooth, thin, orange-yellow skin. The orange, musky-smelling flesh concealed within has a very rich flavour. It is often rather mealy with a consistency of boiled egg-yolk, although there are some better types with a more moist flesh. There are usually between one and three dark brown, shiny seeds present. Canistel is often eaten fresh with a sprinkling of lemon or lime juice, and also adds a nice flavour to drinks and ice-cream. Fruits are nutritious and contain good amounts of carbohydrates, vitamins and minerals. They are believed to have played an important part in the diet of the ancient Peruvians. Other common names used for this fruit include **yellow sapote**, **egg fruit**, **tiesa**, **arbol del hueva** and **huevo vegetal**.

MICROCLIMATE

The canistel is a hardy tree, withstanding light frosts of short duration, and long dry periods with low humidity. It is widely adaptable to most tropical and subtropical areas. Trees are tolerant of a wide range of soils, but prefer the more fertile, free draining types. They have some wind resistance, however sheltered positions are best. Plants grow well in full sunshine.

VARIETIES

Improved selections should bear fruit with a moister flesh. Some varieties planted include: CLEMENTS, AUREA and ROSS SAPOTE.

CULTURE

Plant out in warmer months at least 4 metres from existing trees or buildings. Water in well to displace any air pockets and to settle the soil around the roots. Mulch well to discourage weed growth and help conserve moisture for the shallow surfaced root system. Apply frequent light applications of poultry manure to the mulch as fertiliser.

Water well during long dry periods. The canistel is an easy tree to grow and needs minimal care. Little or no pruning is required other than to provide a good framework with several well spaced branches rather than a central leader.

HARVEST

Seedling trees may take three to five years or more to bear fruit. Grafted varieties are generally less variable in their fruiting habits. Fruit usually mature in late summer months and ripen fully indoors at room temperature in three to twelve days.

PROBLEMS

No serious pests or diseases affect production.

CULINARY USE

The canistel may be eaten fresh with a sprinkling of lemon or lime juice. It is also tasty in milk shakes, custards, pies, sauces, fruit cups, ice-creams and sherbets.

CARAMBOLA

(*Averrhoa carambola*)

From the warm tropical lowlands of Malaysia and Indonesia comes the carambola, *Averrhoa carambola*, also commonly referred to as **star fruit** or **five corners**. This highly ornamental evergreen produces large quantities of succulent, star-shaped fruit. They are a common sight in the market places of South-East Asia where they are popularly eaten fresh or processed into juice.

Carambolas may be sliced up into attractive star-shaped segments, making them an ideal addition to summer salads. The skin is smooth and waxy and may be yellow, orange, or green depending upon variety. In better types the flesh is crisp and juicy and has a sweet, refreshing taste, especially when chilled. There are usually between two and six thin, brown seeds present in each fruit.

The carambola is a must for all home gardens with its delicate light green foliage, small pink flowers, and prolific golden-yellow fruits. It only grows to be a small tree some 5 to 12 metres high.

MICROCLIMATE

Trees prefer a tropical or warm subtropical climate. Young plants may be killed or badly damaged by frosts, while mature specimens can withstand $-2°C$ to $-3°C$ temperatures for short periods with some damage to branches and leaves. The carambola will grow on a wide range of soils from sands to clay loams, however trees prefer well drained clay loams of moderate acidity. They can tolerate dry periods and some wind as long as it is not cold. A warm position in full sun is ideal.

VARIETIES

Seedling varieties should crop in three to eight years, selected grafted varieties in only one or two years. Seedling trees are not generally recommended as they commonly do not produce true to type. Some popular varieties include: FWANG TUNG, ARKIN, B6, B10, B16, MAHA and THAI KNIGHT.

CULTURE

Plant out 4 or 5 metres from existing trees or buildings. In more exposed gardens, individual climate shelters may be required as young plants are very tender. In later years, shelter in these areas can be provided by permanent windbreaks such as hedges. These will reduce the risk of limb breakage, fruit rub and fruit drop. Provide a regular, light mulch and mix in good quantities of well rotted animal manure on a regular basis. Remove lower laterals to 1 metre above the ground as fruit on longer, more pendulous branches may be damaged by touching the ground. Little pruning is required other than to encourage lateral rather than vertical growth for ease of harvesting. Water well during the growing season from fruit set onwards, but reduce quantities just prior to flowering. It is a good idea to thin fruit from large bunches.

HARVEST

The main flowering period occurs in summer, however trees may flower several times throughout the year. Fruit are harvested mainly in mid-summer and early winter, with some fruit maturing at odd times throughout the year. Good crops are harvested from grafted varieties when they are two or three years old, and yields of up to 900 kilograms per year are not uncommon from ten year old trees. A slight colour change heralds maturity, however it is a good idea to pick a few fruits and sample for sweetness. Be careful not to bruise fruit edges as they discolour rapidly. Fruit can be stored in the refrigerator and will keep for ten to twenty days. Freezing fruit is not recommended.

PROBLEMS

Main pests include the fruit fly, fruit sucking moths, and fruit spotting bugs. No serious diseases affect production.

CULINARY USE

Fruits are very refreshing eaten fresh, mixed with other fruits in salads, or processed into drinks. They are also stewed, pickled and used for chutney and jam. The fruit flavour is enhanced by peeling off the 'wing' edges (this removes most of the oxalic acid).

CASIMIROA

(*Casimiroa edulis*)

The casimiroa or white sapote, *Casimiroa edulis*, is indigenous to the highlands of Mexico and Central America. The Aztecs referred to it as **cochiztzapotl** (sleep producing sapote), and the Mexicans know it as **zapote blanco** (white sapote). The seeds, bark, and leaves contain a glucoside, casimirosine, which is used to dilate blood vessels and reduce blood pressure. In large doses the autonomic nervous system is thought to be affected and this helps induce sleep.

Casimiroa is a medium sized, evergreen, open-growing tree which attains a height and spread of about 10 metres at maturity. Tree growth is vigorous. Small whitish-green flowers bloom in spring and produce round to oblong-shaped fruit in summer and autumn. Fruit flesh may be yellow, creamy-white, or orange. It is rich, sweet and melting with a delicious flavour reminiscent of custard apples or premium quality pears. There are usually between two and five large seeds per fruit. A mature tree grown under favourable conditions is capable of producing prolific quantities of fruit every year.

MICROCLIMATE

Casimiroa is a hardy subtropical tree, withstanding drought and both high and fairly low temperatures (down to about −5°C). Frequent light frosts may result in some minor leaf fall. Soils should preferably be fertile and well drained, although trees grow satisfactorily on a wide range of soil types as long as they are not subject to waterlogging. A sunny, sheltered growing spot is best.

VARIETIES

Grafted trees should fruit within two or three years under good growing conditions. Some varieties include: CHESTNUT, LEMON GOLD, SUEBELLE, GOLDEN GLOBE, PIKE, LUKE, SUNRISE, REINECKE COMMERCIAL, VERNON, WILSON, ORTEGO, DENZLER, McDILL, BLUMENTHAL, VISTA, CHAPMAN.

CULTURE

Plant out in warmer months at least 4 or 5 metres from existing trees, and at a greater distance from buildings due to casimiroa's vigorous root system. Water in well to settle the soil around the roots and mulch with compost and animal manure. Try and find a warm, sunny position sheltered from strong winds. Fertilise trees with regular applications of poultry manure added to the mulch. Water trees frequently during the growing season, but reduce the amount during dormancy in winter months. Tip-prune young trees when they are about 60 centimetres high to promote the formation of three main lateral branches. Young trees may need some frost protection for the first two years.

HARVEST

Seasonal varieties are harvested from summer through to autumn months, depending on locality and variety. Everbearing types flower and fruit throughout the year. Immature fruit is green in colour and in several varieties this changes to light yellow when ripe. Hand harvest fruit carefully from the tree while still firm. This fruit should ripen in three to fourteen days (depending upon variety), and may be eaten fresh, cool-stored in the refrigerator, or frozen for future use.

PROBLEMS

Pests include the fruit fly, caterpillars, mites, mealy bugs and scale insects. No diseases seriously affect production.

CULINARY USE

Casimiroa is eaten fresh out of the hand or used in milk shakes, sherbets, ices, ice-creams, and fruit salads. A delicious dessert dish is made by slicing up some chilled fruit and serving it with thickened cream or ice-cream.

CHERIMOYA

(*Annona cherimola*)

One of the finest fruits of the Annona genus is the cherimoya, *Annona cherimola*. The name cherimoya comes from the Peruvian *chirimuya*, which means 'cold seeds'. A close relative of the atemoya, the cherimoya is a small (5 to 7 metres), handsome, semi-deciduous fruit tree native to the highlands of Peru and Ecuador. It is a rather lush looking home garden specimen with its dense cover of attractive, oval-shaped leaves.

The pale-green fruits are round to heart-shaped with a bumpy or corrugated surface that is typical of the Annonas. The white, creamy-textured flesh has a sweet taste and a delicate and delicious flavour, pleasantly aromatic, with numerous brown-black seeds scattered throughout. This sumptuous fruit is considered by many to have no equal and has been described by some as a 'masterpiece of Nature'.

MICROCLIMATE

The cherimoya grows well in a subtropical climate. Being one of the most cold hardy of the Annonas, it can tolerate some low temperatures when mature, however prefers a warm, sunny, frost free growing site. Shelter from strong winds is important as the large leaves tend to act as sails and limb breakage is common. Trees prefer high humidity conditions during flowering for good fruit set. They will grow on a wide range of soils, providing they are deep and well drained in order to discourage root rot diseases.

VARIETIES

Grafted varieties should bear in three to four years from planting out. Some varieties include: WHITE, BAYS, PIERCE, BALWIN, DELICIOSA, SPAIN and CHAFFEY.

CULTURE

Plant out in the warmer months, preferably in early spring after the last of the frosts, and before young trees have emerged from winter dormancy. Plant at least 5 metres from existing trees and buildings. It is a good idea to incorporate large quantities of animal manure, 0.5 kilogram superphosphate and 0.5 kilogram dolomite into the planting site at least three months before planting out. Mulch well and water frequently from bud break through to harvest in autumn or winter. Stake young trees or construct individual climate shelters. No artificial fertiliser should be given in the first year. From the second year onwards apply a NPK (15.4.12.) mix at a rate of 330 grams per year of tree age per annum. Apply this in three equal applications in December, February, and April. Minimum pruning is required in the first year other than to maintain a single leader to a height of 65 or 70 centimetres, then to remove the tip and two or three leaves to promote lateral branching. Be careful to remove any lateral branches that tend to form sharp v-angles to the main trunk as these are prone to splitting. In later years, spring prune back any long, whippy branches and dense crowns to enhance internal bud development.

HARVEST

Fruit matures from April through to August months depending on locality. They are picked firm from the tree and should ripen at room temperature in four to seven days. A yellowing of the space between some of the corrugations on the fruit generally heralds harvest time. A good time to eat them is about two days after they have softened. Fruits don't store well for long periods in the refrigerator and are best kept indoors in a cool spot.

PROBLEMS

Pests include the fruit fly, mealy bugs, scale insects, leaf eating caterpillars and birds. Diseases include root and collar rots.

CULINARY USE

Eat cherimoyas fresh, by themselves, or with other fruit in refreshing salads. Process into drinks, purees, sherbets, ice-cream and sauces.

DURIAN

(*Durio zibethinus*)

From the lush tropical rainforests of Malaysia and Indonesia comes the durian, *Durio zibethinus*, the 'King of Malaysian fruit'. Durian is a medium to large, attractive evergreen, occasionally found growing to a height of 40 to 45 metres in its native forest habitat, however more commonly seen as a grafted orchard tree reaching a height of some 10 to 15 metres.

The durian is adored by many people in its Asian homeland, some are willing to walk miles to roadside stalls in order to obtain these freshly fallen delights. Night long feasts attended by relatives and friends from nearby cities are not uncommon during the harvest season.

The dull olive-green skin of the durian is covered in sharp, stout spines, making it a somewhat formidable opponent to those attempting to procure the deliciously rich, custardy flesh within. The fruit shell can emit a rather strong, fruity odour, however some varieties are milder than others. Each fruit has several large brown seeds embedded in the creamy-yellow pulp.

MICROCLIMATE

The durian is ideally suited to a warm-wet tropical climate with rainfall evenly distributed throughout the year. Trees usually won't tolerate dry periods of three months or more without suffering some damage. A frost free microclimate is important as durians may succumb at any age. Some leaf drop may occur at temperatures less than 8°C, and young trees may die at 3°C and below. Young trees benefit from partial shade until they become established. Good soil drainage is required as trees are susceptible to *Phytophthora palmivora*, and other fungi. Deep sandy to clay loams are preferred. Avoid windy locations unless windbreaks are formed or climate shelters built.

VARIETIES

Seedling durians are unpredictable and may take up to ten years to fruit. Grafted varieties should only take between four and six years, or less, to bear, and have been selected for their superior fruit quality. Some selections include: MONTONG, GAAN YAOW, MAS, BAKUL, D2, D8, D16, TAN CHYE SIAM, SEE PA KU, KK8, SITEBEL, CHOMPOO and CHANEE.

CULTURE

Plant out 6 or 7 metres from existing trees or buildings. In lower lying gardens it may be necessary to mound up planting sites to improve drainage in the root zone. Dig in large quantities of chicken manure and compost to the planting site several months before planting out. The planting hole should be dug just large enough to accommodate the root system comfortably, and plants should be watered in well. Young trees benefit from 30-50% shade until they reach about 1 metre, and thereafter should gradually be introduced to full sun.

Individual climate shelters with some shade cloth are useful. Mulch trees lightly on a regular basis and apply animal manures for fertiliser. Water well during the growing season, but only sparingly during winter to induce prolific flowering in spring and early summer months. Prune dominant upright laterals in order to maintain one central leader.

HARVEST

In Thailand, fruit is cut from the tree and has a shelf life of up to eight days. Malaysians and Indonesians wait until the fruits fall. These mature fruits should be eaten in two to four days. Fruit can be prised open at the spot where the locules (fault lines) come together. Once opened, ripe fruit should be eaten straight away as it rapidly acquires a sour taste. Mature fruit can be stored for up to fourteen days at 10-15°C. Fruit may also be frozen for two to three months.

PROBLEMS

No serious pests or diseases affect production. Rhyparida beetles are the main pest.

CULINARY USE

Durians are usually eaten fresh by themselves or with a sticky, glutinous rice steamed in coconut milk and sugar. Tempoyak is a good way of using quantities of lower quality durian. It is made by frying the flesh until it turns brown and serving it as a side dish with a meal. It keeps quite well, especially if refrigerated. Dried durian conserve (or 'durian cake') is another popular method of preparing lower quality fruits. Cakes and ice-creams are also popular. Durian seeds can be roasted or boiled and have a chestnut flavour.

FEIJOA

(*Feijoa sellowiana*)

The feijoa, *Feijoa sellowiana*, was discovered growing naturally in South America by the German explorer, Sellow, in the early nineteenth century. The tree is thought to have originated in Paraguay, southern Brazil, Uruguay and northern Argentina. It is sometimes referred to as the **pineapple guava**.

The feijoa is a small evergreen tree or shrub with a bushy, spreading habit, and grows to a height of 4 metres or more. It produces an attractive burst of red flowers in spring months. Leaves are coloured green on the upper surface and silvery grey underneath. The fruit is oval in shape with a dark green, or yellowish-green skin and a creamy-white, crunchy-smooth flesh. Feijoas have a most agreeable pineapple-strawberry flavour and a very exotic, pleasant aroma. Like guavas, feijoa fruit contain good quantities of vitamin C. Not only does the feijoa bear tasty fruit, it is also an excellent ornamental shrub and is very useful when grown as a hedge.

MICROCLIMATE

Feijoas are one of the most hardy fruit trees, capable of withstanding both drought and very cold temperatures (as low as −11°C). They are tolerant of a wide range of soil types, providing they are fairly well drained and slightly acidic. When close planted in a hedgerow they are fairly wind resistant and make a handy windbreak. Feijoas are not particularly suited to the more tropical areas in Northern Australia.

VARIETIES

Grafted plants should begin to crop in two to three years from planting out. Seedlings are too variable, and in some instances prove totally unfruitful. Some of the main varieties include: MAMMOTH, TRIUMPH, LARGE OVAL, E4, CHAPMAN and COOLIDGE.

CULTURE

Plant out in warmer months, preferably during periods of expected good rainfall. Allow a distance of about 2 metres from buildings, fences or other shrubs. For hedge planting, a spacing of 1 or 2 metres is recommended. Mulch well with a mix containing large quantities of poultry manure. Spread this evenly under the tree canopy. Fertilise each year with 0.5 kilogram of citrus fertiliser per tree for each year of age. Water trees well during warmer months, particularly during flowering and fruit development. Prune young plants back to a single leader and remove any suckers that appear below the graft. Lateral branches should be pruned annually to avoid overcrowding.

HARVEST

Fruit matures in March to early June following flowering during summer months. Ripe fruit fall to the ground naturally and are collected from the undersurface. Hessian or similar material may be placed on the ground to aid collection. Ground harvested fruit will store well in coolroom conditions for up to six weeks, and fruit picked from the tree may be stored at 3-5°C for two to three months, then allowed to ripen over a period of seven to fourteen days at room temperature.

PROBLEMS

No serious diseases limit feijoa production. The main pest is the fruit fly. Other pests include the scale insect, light brown apple moth, bag moth and macadamia nut borer.

CULINARY USE

Fruit are eaten fresh by themselves, or served with ice-cream or fruit salad. Juice from the fruit is best blended with other fruit juices as it is a bit strong on its own. Feijoa is also used in jellies and preserves.

GRUMICHAMA

(*Eugenia brasiliensis*)

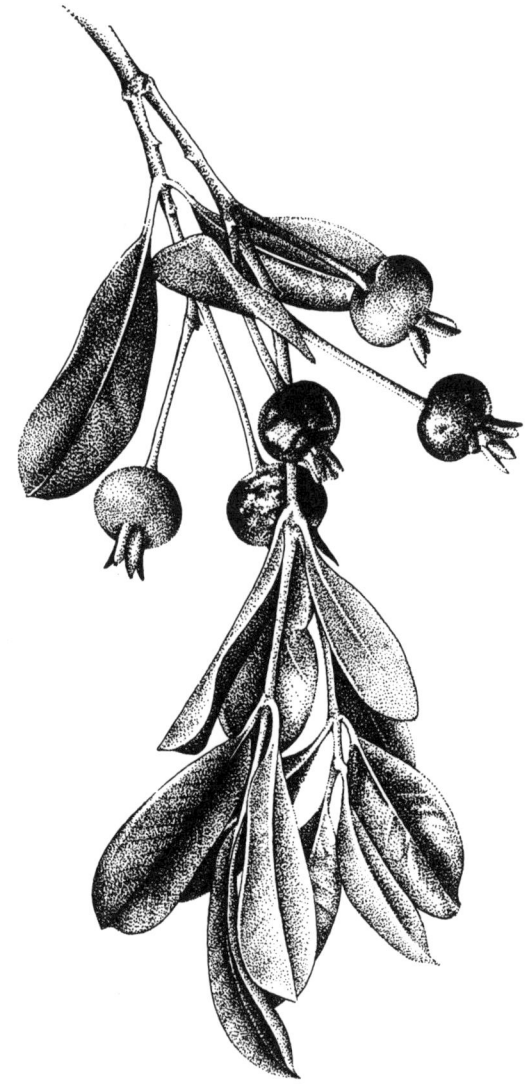

The grumichama, *Eugenia brasiliensis*, makes a fine home garden specimen, valued almost as much for its aesthetic elegance as it is for its sweet, cherrylike fruits. In its native Brazilian habitat, grumichama is sometimes seen as a large tree reaching heights of up to 15 metres, however in most Australian home orchards it grows more slowly to be a small, bushy tree some 6 or 7 metres high.

It is a very handsome tree with glossy, deep green leaves, small white flowers, and reddish young growth. Only four or five weeks after flowering the round, purple-black fruits are ready to harvest. The flesh of the grumichama is soft, melting, and sweet like a cherry. The thin, delicate skin is also edible. Grumichama is also known as **Brazil cherry**, and belongs to the same family as the pitomba, *E. luschnathiana*, and the pitanga, *E. pitanga*.

MICROCLIMATE

Trees perform satisfactorily in a tropical or subtropical climate, withstanding freeze temperatures as low as $-3°C$ for short periods without serious damage. Soils should be slightly acidic, moist, and contain plenty of organic matter. Drought tolerance is low. Shelter from strong winds is advisable. A sunny to semi-shaded position is preferred. Young plants benefit from shade until they become established.

VARIETIES

Seedling trees should only take three to five years, or less, to bear fruit. There are no named varieties that can be recommended at this stage. Variation includes pink or white flowering types.

CULTURE

Plant out in warmer months at least 2 metres from existing trees or buildings. Water in well to settle the soil around the roots, and mulch. Apply frequent dressings of poultry manure as fertiliser. Keep the soil moist at all times, but don't overwater as you may drown the plants. Little or no pruning is necessary as trees grow compactly and maintain an attractive shape.

HARVEST

Trees flower in winter and spring months and the fruit is ready to harvest some four or five weeks later. In tropical locations trees may crop all year. Small clusters of fruit hang down on long stems. They colour from green to red to purplish-black when mature.

PROBLEMS

No serious pests or diseases affect production.

CULINARY USE

Grumichamas are eaten fresh out of the hand or used to make cakes, jellies, jams, pies and liqueurs.

GUANABANA

(*Annona muricata*)

From the American tropics comes guanabana, *Annona muricata*, also commonly known as the **soursop**. Guanabana is a small (6 to 9 metres), slender, upright growing evergreen with smooth, shiny, dark green leaves that are aromatic when crushed. They are found growing wild throughout the West Indies and are very popular orchard trees in Cuba, Colombia, Puerto Rico, and the Bahamas.

Fruits are the largest of the Annonas, each weighing at least 1 or 2 kilograms. They are oval, heart, or kidney-shaped and when mature have a yellow-green, leathery skin with a covering of soft, fleshy spines. The white, pulpy flesh inside is pleasantly aromatic, very juicy and has a slightly acid taste. Refreshing drinks processed from guanabana are popular throughout the American tropics. The rich, creamy juice, when mixed with a little milk, makes the fabulous 'champola'.

Other names for guanabana include **corossol, graviola, guayabano, sorsaka** and **zuurzak.**

MICROCLIMATE

A warm, moderately humid climate is preferred. Trees of all ages are frost sensitive and may be damaged at temperatures less than 0°C. A sunny position sheltered from strong winds is desirable. Soils should ideally be rich, deep, and well drained. Mulching is recommended to avoid dehydration of the plant's shallow, fibrous root system.

VARIETIES

Sweeter varieties are more suitable for home garden plantings. The acid types are normally only used for processing into drinks. Grafted varieties should bear in about three years, seedlings from three to five years. Varieties include the CUBAN FIBRELESS.

CULTURE

Plant out at least 2 or 3 metres from existing trees or buildings. Mulch well around the base, however not too close to the trunk or you may predispose trees to disease. Provide some wind shelter in more exposed gardens. Apply poultry manure to the mulch as fertiliser every three months to encourage good growth and fruiting. Water regularly, particularly during spring and early summer months. Minimal pruning is required other than to trim back any long, whippy branches.

HARVEST

Harvesting is carried out most of the year, the main season being January to May. Fruit are picked when full grown and still firm, and slightly yellow-green in colour. Handle with care to avoid bruising. They will ripen in a few days at room temperature, and may be eaten when the flesh is slightly soft. Ripe fruit should keep in the refrigerator for several days.

PROBLEMS

Young trees are very cold sensitive. No serious pests or diseases affect production. Some trees may suffer die-back due to root problems.

CULINARY USE

Sweet fibre-free varieties are often eaten fresh by themselves or in fruit salads. Refreshing drinks and sherbets are made from the rich, creamy juice. The juice is obtained by pushing the white, pulpy flesh through a colander or squeezing it in cheesecloth. Guanabana is delicious when whipped with ice-cream, or made into custards, mousses, jellies and souffles. Sun dried pieces are a real treat. The seeds are toxic, so make sure that you remove them all from the pulp before processing.

GUAVA

(*Psidium guajava*)

The guava is native to the American tropics and was found growing naturally from Peru to Colombia by the Spanish conquistadores. The name is thought to be derived from the Haitian name for the fruit, guayaba. The Spanish and Portuguese were responsible for its spread throughout the world.

There are two main species grown: the **common guava**, *Psidium guajava*, and the **cherry guava**, *P. cattleianum*. The cherry guava produces a red, acid fruit preferred for processing into drinks and purees, or jams and jellies. The common guava is relatively sweeter, with a yellow, white, or pink flesh and the better types are used as dessert fruit. Both species are recognized as a valuable source of vitamin C.

The guava grows into a small, evergreen spreading tree or, more commonly, a large, shallow rooted shrub growing to between 3 and 10 metres. The greenish or reddish-brown bark is smooth and flaky.

Abius. (D. Cilento/QTTC)

Flower of the Malay apple.
(B. Scomazzon)

Jaboticabas. (J. Truscott)

Miracle fruit. (D. Cilento/QTTC)

Grumichamas. (D. Cilento/QTTC)

Akee. (D. Cilento/QTTC)

Mangos. (J. Truscott)

Market — rambutans and
mangosteens. (R. Francis)

MICROCLIMATE

Guavas are very hardy and grow well with little attention in a wide range of soils and climates. High temperatures and drought conditions are usually tolerated. Mature dormant phase trees should survive light frosts, however young trees may die under similar conditions. Guavas grow best in warm, sheltered, frost-free locations and respond well to high temperatures during fruit development. They generally tolerate short periods of waterlogging.

VARIETIES

Guavas grown from seed normally don't reproduce clonally from their parents and may produce inferior fruit. Improved varieties suitable for home gardens are produced by propagation methods including cuttings, grafting, budding and marcotting. These should bear fruit in less than eighteen months from planting out. Some varieties include: ALLAHABAD SAFEEDA, LUCKNOW 49, HONG KONG PINK, RUBY X SUPREME, MALHERBE SAXON, VAN ZYL, FAN RETIEF.

CULTURE

Guavas may be planted out at any time during the warmer months, provided that young plants are watered frequently during dry weather. Individual plant shelters may be required in more wind prone gardens while more permanent windbreaks are being formed. Plantings should be made at a distance of 2 or 3 metres from existing trees or buildings. Mulch well to help conserve soil moisture and apply 0.5 kilograms of a complete artificial fertiliser in early spring every year. Prune lower branches from six month old plants, and in general promote multiple branching by cutting back long apical branches. Remove any dead plant matter and crowded or crossed branches.

HARVEST

Guava trees usually flower in spring or early summer following a period of dormancy. Fruit normally matures in late summer to autumn. In warmer climates guavas may fruit almost continuously throughout the year. Ripe fruit may be stored satisfactorily in the refrigerator for up to two weeks.

PROBLEMS

One of the main pests is the fruit fly. Anthracnose disease may be a potential problem, together with various fruit rots.

CULINARY USE

Guava fruit is eaten raw, made into refreshing fruit juices, jams, jellies, toppings, syrups, purees, marmalades and milk shakes.

JABOTICABA

(*Myrciaria cauliflora*)

The jaboticaba, *Myrciaria cauliflora*, bears large quantities of purple-black, grape-like fruit in a spectacular fashion directly onto the trunk and main branches of the bush, only twenty to thirty days after flowering. The fruit possesses a thin, tough, edible skin and has a very sweet, slightly aromatic, milky-white, translucent pulp with a pleasant grape flavour. Trees may crop several times from spring through to autumn months. There is a giant and a dwarf tree type. The giant type bears larger fruit, often with several seeds.

The jaboticaba is a small, slow growing, bushy evergreen tree, ideal for the home garden. It is occasionally found growing to a height of 10 to 12 metres in its native southern Brazil, however is more commonly seen in Australian home gardens as a large bush growing up to 5 metres high and 3 metres in diameter. It grows compactly with a dense cover of small, pale-green leaflets. New growth flushes, in warmer months, are a very decorative reddish-pink colour.

MICROCLIMATE

Trees perform best in a warm, sunny position protected from strong winds. They grow very slowly unless well nourished in rich soils with a slightly acid pH. Good growth is required to maintain fruit size and quality. The jaboticaba should have constant soil moisture as permanent damage may occur during long, dry spells. Light frost is tolerated, however growth stops during prolonged cold periods. Plants have a low tolerance to salt.

VARIETIES

Seedlings are variable and very slow growing when young. They may take up to five or six years to bear. Grafted or air layered plants should only take four years to fruit. Some selections include: J1, YOUNGHANS and SABARA.

CULTURE

Plant out in a sunny, well drained and sheltered position at least 2 metres from existing trees or buildings. Mulch well with a layer of compost and poultry manure, however not too close to the trunk or stem rot may occur. Water well during the warmer months of the year. In more exposed gardens it may be necessary to grow windbreak hedges. Jaboticaba is a slow grower and needs plenty of tender loving care to maintain good growth. Little or no pruning is required.

HARVEST

Small white flowers bloom profusely on the trunk and main branches in several bursts from spring through to autumn. Only twenty or thirty days later fruit is ready to harvest. Fruits store well in the refrigerator for one or two weeks.

PROBLEMS

No diseases seriously affect production. The fruit spotting bug can be an occasional pest.

CULINARY USE

Jaboticaba may be eaten fresh out of the hand or used in jams, jellies, syrup, cakes and juices. An excellent wine is also made from the fruit.

JAKFRUIT

(*Artocarpus heterophyllus*)

A very handsome, quick growing evergreen, reaching a height of 8 to 20 metres, and bearing the world's largest fruit in a most unusual manner from the trunk and main branches. This is the jakfruit, *Artocarpus heterophyllus*. Thought to be native to the rainforests of the Western Ghats (mountains) in India, it is now commonly grown in most tropical lowland regions of South-East Asia. Jakfruit and other starch plants, such as sweet potato, cassava and breadfruit, are important food sources for the local inhabitants.

The jakfruit tree is a fine looking specimen with lush, glossy, dark green leaves and is very decorative in the home garden. Grafted trees bear fruit in about two to four years from planting out, whereas trees grown from seed may take between three and eight years.

Giant-sized fruits weighing up to 46 kilograms or more and containing several hundred seeds are not uncommon. The yellow or brown rind is covered in short, hard spines. The golden yellow to pink flesh within is sweet and juicy.

MICROCLIMATE

Trees thrive in the hot, humid lowland tropics, however they adapt well to a wide range of climatic conditions including the subtropics, and will survive light frosts when mature. There should be an even distribution of rainfall throughout the year. Soils should preferably be deep and well drained, and of a sandy or clay loam structure. Soil moisture should always be kept at a high level, but stagnation is not tolerated.

VARIETIES

Propagation of varieties is often by seed, so there can be a wide variation in fruit characteristics from tree to tree. There are two main types recognised; those with a firm, crisp pulp and those with a soft and melting pulp. Some selected varieties include: BOSWORTH, GALAXY, FITZROY, NS No. 1, NAHEN, CHEENA X, KAPA, MUTTON VARIKKHA and SINGAPORE.

CULTURE

Plant out in the warmer months at least 5 or 6 metres from existing trees or buildings. Dig the planting hole just large enough to accommodate the root system comfortably. Be careful not to damage the roots while planting, and water the soil well to displace harmful air pockets. Some shadecloth protection for the first few weeks is a good idea until the young plants establish themselves. Mulch on a regular basis with compost. Apply some animal manure to the mulch with each application. Young trees will probably need some regular watering during the first year or so, but after this they can be neglected except in very dry periods. Little or no pruning is required other than to remove any dead branches from the interior of the tree, so that sufficient light is obtained for the developing fruit (fruit grow on stems originating from the trunk and main limbs).

HARVEST

Harvesting is normally carried out from January to June depending on locality, however trees may bear fruit throughout the year. When mature, there is usually a change of fruit colour from light green to yellow-brown. Spines, closely spaced, yield to moderate pressure, and there is a dull, hollow sound when the fruit is tapped. Another sure sign of maturity is when the fruit starts to emit its characteristic odour.

PROBLEMS

Some of the main problems include aphids, scale, mealy bugs, grasshoppers, wood borers, fruit sucking bugs, and pink's disease.

CULINARY USE

Immature fruit is used as a vegetable and may be fried, roasted, or boiled. Ripe fruit is usually eaten fresh as a dessert fruit. The fleshy segment that encloses the seed is the choicest part of the fruit, however the 'rags' or stringy flesh in between is also quite tasty. The seeds are boiled or roasted and in some countries are more highly valued than the pulp itself.

KIWIFRUIT

(*Actinidia deliciosa*)

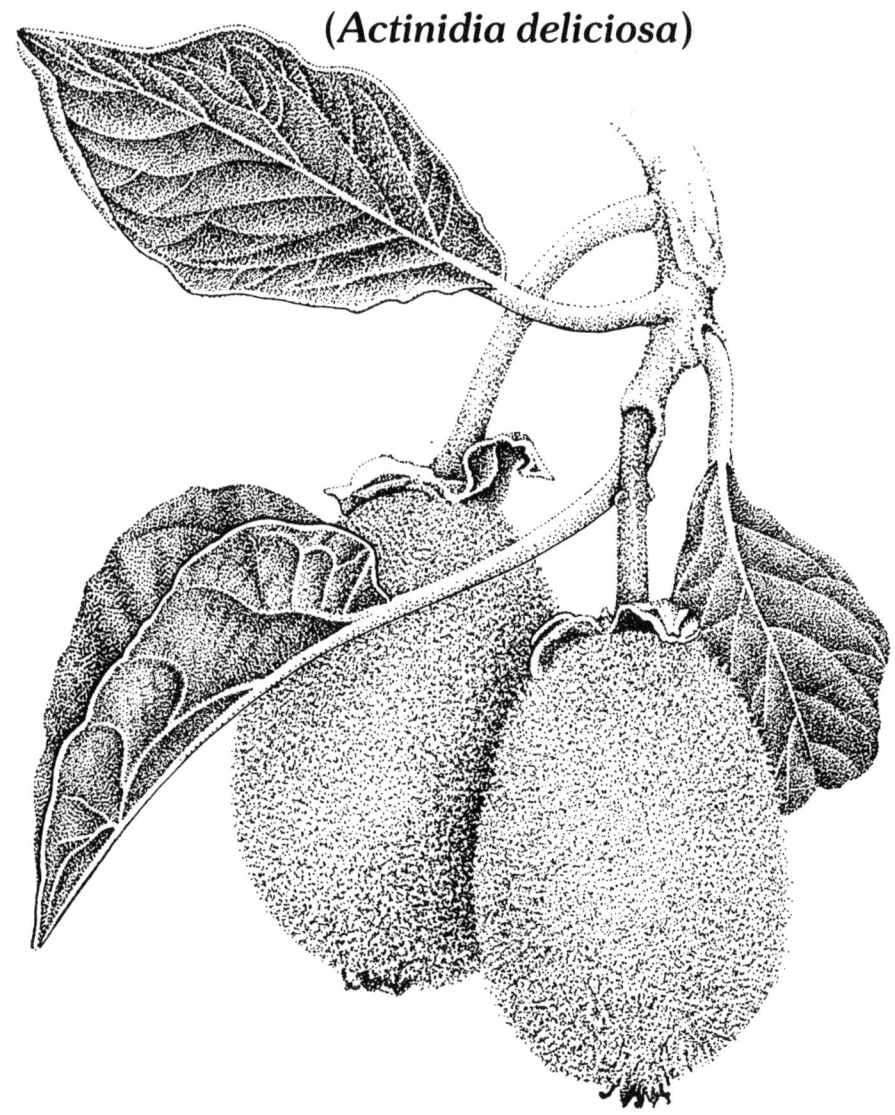

The irrepressible kiwifruit is oval-shaped and has a firm, brown, hairy skin. The flesh is a very attractive light green with a regular pattern of numerous, small dark seeds radiating from the centre. Fruits possess a sweet, delicate flavour, contain large quantities of vitamin C, and are popular eaten fresh or in dessert dishes, such as the delicious pavlova.

The kiwifruit, *Actinidia deliciosa*, is a large, vigorous, deciduous fruit vine which is found growing naturally to a height of 9 metres or more on the forest margins in the Yangtze Valley, China. New Zealand was one of the first countries to produce this increasingly popular and delectable fruit on a commercial basis and the industry presently boasts a very high level of technology and marketing expertise. The majority of plantations in Australia are concentrated along the eastern seaboard from Maleny, Queensland, south to the Dandenong Ranges in Victoria. The kiwifruit is also referred to as the **Chinese gooseberry** and **yangtao**. The vines are very long-lived and orchards normally produce good crops for twenty to thirty years.

MICROCLIMATE

In southern Australia plantings should preferably be made in locations with warmer, north or easterly aspects, free from heavy spring frosts. In northern areas, in mild seaside locations, a cooler southerly aspect may prove more productive than a northerly one. As the kiwifruit is a deciduous vine, winter frosts rarely cause any damage. Kiwifruit will grow on a wide range of soils, provided that they are relatively deep and well drained. Plants require protection from strong prevailing winds, in order to prevent tender young laterals being snapped off at their bases and fruit being blemished by wind rub. In inland areas, vines should be protected from hot, drying winds.

VARIETIES

Kiwifruit vines are dioecious with separate male and female plants required for fruit production. Some female fruiting varieties include: DEXTER, HAYWARD, BRUNO, ABBOTT, MONTY, ELMSWOOD, FUSARELLI and GRACIE. Double grafted kiwifruit vines are well suited to the smaller home gardens. Seedlings may take between four and seven years to bear, whereas grafted vines usually take only one and a half to three years.

CULTURE

Plant out in the warmer months of the year. Support structures such as trellises or pergolas will be required. One male vine is sufficient to pollinate five to eight females. Mulch around each vine and mix in some well rotted animal manure on a regular basis as fertiliser. Water regularly during the growing season as kiwifruit vines have shallow, fibrous root systems. Train vines to grow with a single straight trunk to the top of the trellis or pergola, and a single strong leader along the central overhead wire or cross-bar in each direction. Fruiting arms (25 to 40 centimetres apart) are trained at right angles to the leaders and tied down carefully to the outside supports. The first crop of fruit forms on these arms, and later crops form on laterals that develop from them. Kiwifruit crop on current season fruiting laterals that originate from one year old fruiting arms. Fruit generally develops on the bottom four to six buds of each lateral. Laterals are replaced every four to six years, before they become overlong, by pruning back to permanent fruiting arms. Males can be pruned heavily just after flowering each year.

HARVEST

Harvesting is carried out between April and June, depending upon variety. Fruit are harvested in a firm, unripe condition and allowed to soften up at room temperature. Kiwifruit can be cool stored at 0°C for up to four months, providing that it is picked firm at the beginning of the harvest season. Without refrigeration fruit will store in a cool spot for up to eight weeks.

PROBLEMS

Pests include soil nematodes, pasture white grubs and various leaf rollers, beetles and caterpillars. Diseases may be due to *Botrytis*, *Diothorella*, and a stem canker caused by *Phytophthora* or *Pythium sp.*

CULINARY USE

Fruits are eaten fresh by themselves or used in salads, fruit creams, ice-creams, mousses, cakes, sorbets, jams, pickles and chutney.

LANGSAT AND DUKU

(*Aglaia domesticum*)

The langsat and duku, *Aglaia domesticum*, are thought to be native to Malaysia, Indonesia, India, Burma, and the Philippines. They are very popular fruits in South-East Asia, and have been in cultivation for many years. Both langsat and duku are classified as a single species, however they differ in growth habit and fruit characteristics.

The langsat is a tall (10 to 20 metres), slender evergreen with a rather straggly crown. Fruits have a sweet or sour, white translucent flesh. They are oval-shaped with thin yellow skins containing a milky latex, and are produced in bunches of fifteen to twenty-five. Duku trees are generally smaller with an attractive rounded shape. Duku fruits are usually larger than langsat, more round in shape, have a sweeter flesh, a thick latex-free skin, and are produced in bunches of between four and twelve. Better quality fruits from both trees have a refreshing, rather aromatic flesh.

MICROCLIMATE

Langsat and duku are tropical trees, however with a little extra care and attention they should grow in warmer subtropical areas. Both trees tolerate high temperatures up to 40°C, however low temperatures usually result in damage. Light frosts may defoliate plants, often leading to severe die-back in young trees. Mature trees usually regenerate in spring months. Trees grow successfully on a wide range of soils, preferring a free draining sand or clay loam, high in organic matter and moist (particularly during flowering and early fruit development). Young trees benefit from partial shade until they are 2 or 3 metres tall. A warm, humid, sheltered position is best.

VARIETIES

Seedlings are very slow to come into bearing, often taking up to fifteen or twenty years. Grafted trees usually take six to eight years, some air-layered plants only two to four years. Some varieties include: DUKU-LANGSAT (hybrid), PAETE (langsat), UTTARADIT (langsat), DUKU (duku), JURONG (duku) and LONGHONG (duku).

CULTURE

Plant out in warmer months, especially during periods of expected good rainfall, at least 4 metres from existing trees or buildings. Mulch well and provide at least 50% shade under the canopies of other larger trees, or use shadecloth. Young trees should be gradually introduced to full sunshine when they are about two or three years old (or 2 to 3 metres tall), or they may be left to grow well in partial shade. Apply frequent light applications of poultry manure to the mulch as fertiliser. Water well, especially from fruit set onwards. Little or no pruning is required, other than to head back the leading stem in seedling langsats to provide a more branched structure.

HARVEST

Fruit are normally harvested from January to May months. Skin colour changes from green to yellow at maturity. Bunches ripen at different times, however all fruit in a single bunch usually ripens together. They should keep for up to four days at room temperature.

PROBLEMS

Collar rot disease may be a problem. Few serious pests or diseases affect production.

CULINARY USE

Fruits are eaten fresh out of the hand, used in salads, or preserved.

LIME

(*Citrus aurantifolia/latifolia*)

The lime is native to Southern Asia and is used extensively in the tropics for juices such as limeade and cordials, food flavourings and pickles. It also makes a good marmalade, and lime oil is prepared from the peel. The fruit was believed to be a good source of vitamin C and for this reason it was carried on long sea voyages by British sailors to prevent scurvy (hence the term 'limeys').

The lime tree, *Citrus aurantifolia/latifolia*, is a small, bushy evergreen that grows to a height of about 5 metres when mature. Leaves are light green and oval in shape, emitting a pleasant lime fragrance when crushed. The branchlets and twigs usually boast an impressive armoury of short, sharp spines. Fruits are round to oval with thin, yellow-green skins. The greenish pulp within is juicy, very acid, and is used in drinks rather than eaten fresh.

MICROCLIMATE

Lime trees should be grown in warm, frost free locations. Trees grow best in full sunshine and should be given plenty of protection from cold winds. They are adaptable to many soil types including rocky alkaline soils and sandy acid soils. They are hardy specimens and usually grow satisfactorily even when neglected.

VARIETIES

There are three main types of lime. The **West Indian lime** (*Citrus aurantifolia*) is suited to the hot lowland tropics, and grows well in warmer, frost free locations. The **Tahiti lime** (*Citrus latifolia*) is more cold tolerant, has some disease tolerance, but lacks the flavour and aroma of its more tropical counterpart. The **Rangpur lime** (*Citrus X limonia*) is actually a mandarin-lime hybrid with a tangerine coloured fruit. Propagation of most types is usually done by seed, but in the case of Tahiti lime, which is seedless, it is done by cuttings or grafting. Trees normally take two to three years to bear fruit.

CULTURE

Plant out in warmer months at least 2 metres from existing trees or buildings. Water in well to settle the soil around the roots and to displace any damaging air pockets. Mulch to discourage weed growth and to stimulate soil micro-organism activity. In colder, more exposed gardens, it may be necessary to construct individual climate shelters while permanent windbreaks are being established. Fertilise with regular dressings of poultry manure. Little or no pruning is required other than to trim for symmetry and remove any dead or damaged growth.

HARVEST

Fruits ripen from autumn through to spring months. They will fall from the tree when fully ripe, however they are often harvested before they yellow. Excess fruit will keep in the refrigerator for two to three months, or can be sliced and stored in the freezer.

PROBLEMS

Melanose fungal disease in coastal areas. Also tristeza, anthracnose and phytophthora disease may cause some problems.

CULINARY USE

Due to their rather sour taste, limes are not normally eaten as a dessert fruit. They do make a very refreshing juice though, and are also used in sorbets, pies and marmalades.

LONGAN

(*Euphoria longan*)

Affectionately known as the 'little brother of the lychee', the longan, *Euphoria longan*, is thought to be native from southern China to India. It is extensively cultivated on a commercial basis in Taiwan, northern Thailand, and the Fukien province of China.

The longan makes a fine ornamental with its lush cover of glossy, dark green foliage. It is a handsome, well shaped, round-topped evergreen that requires little pruning. Trees grow to a height of 10 to 13 metres and closely resemble the lychee in shape and appearance.

Longan fruits are round to ovoid in shape, somewhat smaller than the lychee, and are produced in loose clusters of between ten and eighty, or more. They are yellow-brown or greenish-brown when mature and have smooth, tough, leathery skins. The sweet, white, translucent flesh is similar in texture to lychee and rambutan, with a flavour that is a little stronger, muskier and somewhat spicier. Some varieties are soft and juicy, others crisp and crunchy.

MICROCLIMATE

The longan grows well in a subtropical to temperate climate. Mature trees are fairly cold hardy and survive temperatures down to approximately −4°C. They also grow and fruit in warm, northerly regions. Trees usually do best in areas with dry, cool winters followed by high spring and summer temperatures. They grow satisfactorily in most soils, preferring deep, fertile, alluvial clay loams where the water table is high. A good supply of soil moisture is necessary from fruit set through to harvest. Trees stand up well to wind and are sometimes used as windbreaks.

VARIETIES

Seedling longans are not recommended as they frequently don't produce true to type and may take as long as twenty years to bear fruit. Air layered or grafted plants of superior cultivars should produce good quality fruit in three to four years from planting out. Some popular varieties include: DAW, DANG, CHOMPOO, HAEW, BIEW KIEW, BAIDUM, KOHALA, HOMESTEAD, CHIEN LEIU and DUAN YU.

CULTURE

Plant out in warmer months at least 5 metres from existing trees or buildings. Handle young plants with care and avoid carrying the tree by the trunk. Cut the base out of the planting bag and carefully slip the plastic up and over the branches. Each tree should be given 12 to 16 litres of water in order to displace air pockets and to settle the soil around the roots. Mulch well, to help conserve soil moisture and reduce competition from weeds.

In more exposed gardens it may be necessary to provide individual climate shelters constructed with hessian or synthetic material.

Once young trees have become established and produce their first growth flush, fertilisers may be applied. For bearing trees this is done at fruit set and just after harvest. Avoid fertiliser applications prior to flowering time. Fertiliser rates as for lychee. Some regular light applications of poultry manure to the mulch is a good idea.

Pruning should be minimal other than skirting the lower branches that may rub on the ground. Water trees well from fruit set right through to harvest.

HARVEST

Harvesting is usually carried out from January to April depending upon variety and microclimate. Maturity is determined by fruit shape, skin colour and taste. It is a good idea to pick a few fruit and sample for sweetness before all fruit is harvested.

Fruit freeze well and retain their flavour and colour when thawed. They may also be cool stored for several weeks in the refrigerator to extend their shelf life.

PROBLEMS

Pests include birds, flying foxes, and lepidopterous caterpillars.

CULINARY USE

Fruits are usually eaten fresh by themselves or in fruit salads and fruit cocktails. They combine well with banana, pineapple, papaw and grapes. They may also be dried, stewed or canned.

LUCMO

(*Pouteria obovata*)

The lucmo, *Pouteria obovata*, bears large quantities of round to flattish-round fruit varying in size from an apple to a grapefruit. They have a tender, greenish-yellow skin covered in small, white pores and a sweet, dry, mealy orange flesh with between one and three large seeds. The fruit is also known as the **lucuma**, **lugma** or **lucma**. The tree is native to southern Ecuador and the Andean foothills of Peru and northern Chile. It is popular in Chile where it is grown commercially for eating fresh and flavouring ice-cream.

Lucmo is an open-growing, evergreen fruit tree closely related to the canistel, *P. campechiana*, and the abiu, *P. caimito*. Under ideal conditions it grows to a large size and may reach 20 metres in height, however it is more commonly seen in Australian home orchards as a medium tree to 10 metres. Leaves are glossy, dark green, stiff and leathery, and are clustered near the ends of the twigs and smaller branches.

MICROCLIMATE

Lucmo is widely adaptable to most tropical and subtropical areas. In its native, high altitude, Andean valley habitat, the trees grow well where the temperature is warm for a few hours around midday, but cool for the remainder of the day, with little or no frost. The tree will grow in poor soils, but does best in a deep, alluvial, free-draining soil, high in organic matter. The lucmo also grows satisfactorily in a dry climate with low rainfall. A sunny, sheltered position is preferable. Young trees should be given some frost protection.

VARIETIES

No named varieties are recommended at this stage. Seedling trees are variable in growth habit and fruit quality and may take up to eight years to bear.

CULTURE

Plant out in warmer months at least 4 or 5 metres from existing trees or buildings. Water in well and apply a mulch to the soil surface extending under and just beyond the leaf canopy. Avoid placing mulches too close to the base of the trunk. Apply frequent light applications of poultry manure to the mulch as fertiliser. Water during long, hot-dry periods. Little or no pruning is required, other than to provide a good framework with several well spaced branches instead of one central leader. The lucmo is an easy tree to grow and needs minimal care.

HARVEST

Fruit mature in autumn and ripen fully indoors at room temperature in three to twelve days. They may be cool stored in the refrigerator or frozen for later use.

PROBLEMS

No serious pests or diseases affect production.

CULINARY USE

Fruits are eaten fresh with a pinch of salt or a sprinkling of lemon juice. The pulp is used to flavour ice-cream and milk shakes. In Peru it is ground into a meal or flour which adds a strong odour and colour to desserts, ice-creams, sherbets, puddings, punches and milk shakes. Lucmo fruit can also be preserved.

LYCHEE

(*Litchi chinensis*)

The lychee, *Litchi chinensis*, is a most attractive, round-topped evergreen tree possessing fruit with a delightful, sub-acid flavour. The tree is native to a small area in subtropical south-eastern China where it has been grown successfully for many centuries. It is thought to have been first cultivated by people of Malaysian descent before the arrival of the Chinese. The tree was first introduced into Australia by early Chinese gold seekers during the 1870s goldrush era. The major world producers include China, India, Taiwan, South Africa, Thailand, Mauritius, Hawaii, Hong Kong and Australia.

The lychee tree is considered to be a fairly slow grower, however when mature (about twenty-five years) attains a height of 7 to 15 metres and a crown diameter of up to 12 metres. The fruit is ovate in shape and is produced in loose clusters between one and forty. Lychees have a rosy red leathery skin or rind, rather brittle, with a rough surface. Each fruit has a central brown seed surrounded by white, semi-translucent flesh. Lychee fruit is a rich source of vitamin C, contains 10-15% soluble sugars, and also good amounts of phosphorus, niacin, riboflavin and vitamin B2.

MICROCLIMATE

Lychee production is enhanced by a cool, dry, frost free spell during autumn and winter which induces flowering, followed by high spring and summer temperatures and humidity. Persistent wet or windy weather during flowering may result in poor fruit set. Mature lychee trees withstand light frosts, however young plants should be given some protection. A regular, even moisture supply is necessary from flowering through to harvest. Soils should ideally be deep, free draining 'fine' sandy loams, with plenty of organic matter, however trees perform satisfactorily on a wide range of soil types provided that waterlogging or water-stress doesn't occur for extended periods. Avoid windy locations as tree branches are brittle.

VARIETIES

Seedling lychees may take seven to twenty years to bear. Air layered or grafted plants of recognised varieties should fruit in three or four years or less. Early season varieties are more suited to warm, northerly areas, while mid to late season varieties grow further south. Some of the main varieties listed in order of maturity include: TAI SO, HAAK YIP, BENGAL, KWAI MAY (Red and Pink), NO MAI CHEE and WAI CHEE. Other varieties include: BREWSTER, TAI SO-MAURITIUS, SOUEY TUNG, GROFF and MUZAFFARPUR.

CULTURE

Plant out in a warm, sunny position at least 5 metres from existing trees or buildings. Add 0.5 kilogram of blood and bone, then some compost to the bottom of the planting hole. Provide individual climate shelters to protect young trees from wind and sunburn. Mulch well and fertilise young trees regularly (no artificial fertiliser should be given at planting time). Once trees begin fruiting they are fertilised strategically to control vegetative flushes. Bearing trees should be given each year 0.4 kilogram of a well balanced NPK mix per tree for each year of age (apply 40% at fruit set, and 60% after fruiting). Water regularly from fruit set until late summer to early autumn but reduce this amount for a few months prior to flowering. Trees should be encouraged to produce one good vegetative flush just after harvest and then be left in a dormant condition during late autumn and winter before flowering. Pruning should be aimed at forming a sound tree frame work with four or five main branches with wide angles to the trunk. Mature trees need only a light prune just after harvest to encourage flushing.

HARVEST

Fruit is harvested over a two or three week period in early summer (north) to early autumn (south), depending upon varieties and seasonal conditions. They ripen on the tree all about the same time, however the harvest period may be stretched to a couple of weeks. Pick fruit when it is soft to touch, a full red colour, and sweet to taste. Flattening of the tubercles on the rind is normally a good sign of maturity. Harvest fruit by cutting as little of the stem as possible above each fruit cluster. Pick from the crown of the tree downwards over a period of at least a week. Fresh lychees may be stored for two or three weeks at 2-7°C. They can be frozen but this results in a total loss of colour.

PROBLEMS

Pests include the macadamia nut borer, macadamia flower caterpillar, fruit fly, fruit bats, erinose mites. Diseases include bark canker and brown leaf felting.

CULINARY USE

Lychees are best eaten fresh to appreciate their fine flavour. They can be eaten in fruit salads, and are tasty in dishes with pork or chicken. They are also canned and dried.

MABOLO

(*Diospyros discolor*)

From the Philippine Islands comes the mabolo, *Diospyros discolor*, also known as **velvet apple** or **butterfruit**. The tree belongs to the Ebony family which is well known for its hard, black wood, ideal for artistic carvings inlaid with ivory.

Under more favourable conditions in its native habitat the mabolo is a large tree occasionally found growing to a height of 20 metres or more, but more commonly it is a slow growing tree only reaching some 8 or 10 metres in most Australian home gardens. It is quite decorative with its very long, smooth, dark green leaves and graceful, weeping branches.

Fruit are pink to reddish-brown to purple in colour, round to oval-shaped, and have a thick covering of silky, velvety hairs that are usually rubbed off before eating. Better quality fruit has a creamy-white, soft and mealy flesh with a sweet flavour and an aroma slightly resembling cheese. To some people it is an acquired taste, however one that usually soon becomes popular. The seedless types often make better dessert fruits as they are generally more moist and sweet.

MICROCLIMATE

The mabolo prefers a warm to hot, subtropical to tropical climate. Mature trees usually withstand frost for short periods, however some leaves and branches may die. Young plants normally don't tolerate these low temperatures and should be given every protection. Trees of all ages suffer badly in more exposed gardens that are subject to frequent cold winds. Soils should be slightly acidic, moist and free draining. A sunny or partially shaded position is best. Salt tolerance is low.

VARIETIES

Seedless varieties are generally planted in preference to the poorer quality seeded types. Seedlings usually take five or six years to fruit. Selected varieties include: MANILA × EDWIN BELAN.

CULTURE

Plant out in the warmer months of the year at a distance of at least 5 metres from existing trees or buildings. Dig holes just large enough to accommodate the root system comfortably and allow for some lateral growth. Water in well to settle the soil around the roots and apply a mulch to help conserve soil moisture and retard weed growth. Apply frequent light dressings of poultry manure to the mulch as fertiliser, however avoid applications just prior to flowering as fruit yields may be decreased. Little or no pruning is required other than to remove dead or dying branches and inside growing shoots. Water regularly during the growing season, but tend to neglect trees for a short period just before flowering.

HARVEST

Fruit mature in late summer to autumn and are picked when they have attained full size and colour. Pick some fruit and sample for taste before harvesting the whole crop. Fruit are ready to eat when they are soft, like their close relative the persimmon.

PROBLEMS

No serious pests or diseases affect production.

CULINARY USE

Mabolo is eaten fresh or made into fruit juice, sherbets and table jellies.

MACADAMIA

(*Macadamia tetraphylla/integrifolia*)

The macadamia nut, *Macadamia tetraphylla* and *Macadamia integrifolia*, is found growing wild in the subtropical coastal rainforests of southern Queensland and northern NSW in eastern Australia. In its native forest habitat it grows as a seedling with a tall, thin trunk competing for sunlight with other trees at the top of the rainforest canopy. In earlier days, because of land clearing for growing bananas and other small crops, the native macadamia developed into a more symmetrical specimen and is now more commonly seen in Australian orchards and home gardens as a small compact evergreen, growing to between 7 and 10 metres.

Leaves are dark green, clustered in whorls of three or four, and in some types have sharp, spiny edges. The golden-brown, hard shelled nuts are encased in smooth, green, fleshy husks which split open when the nuts ripen. Macadamia nuts are a little hard to crack open, however the effort is well worthwhile, as they contain a delicious creamy-white kernel, now commonly regarded as one of the finest nutmeats in the world.

MICROCLIMATE

Trees are suited to a warm subtropical climate. A frost free position is best as young plants may be damaged or killed at low temperatures. Shelter from cold winds is essential to maintain good growth. Soils should be well drained, contain good quantities of organic matter, and be friable enough to permit root growth to a depth of at least 1 metre. Trees grow and crop best in full sunshine. Salt tolerance is low.

VARIETIES

There are two main edible species of macadamia. *M. tetraphylla* bears a rough, pebbly shelled nut with a sweet flavour that is best eaten fresh. Trees have leaves with prickly, saw-toothed edges. They grow well in cooler, more southern areas. Some varieties include: ELIMBAH, RENOWN, NUTTY GLEN (hybrid). *M. integrifolia* produces smooth shelled nuts that are suitable for roasting. Most commercial orchards grow selected varieties of this species. Some varieties include: HINDE, KAKEA, KEAAU, IKAIKA, KEAUHOU. Grafted trees bear in three or four years or less, however seedling trees may take seven years or more.

CULTURE

Plant out in warmer months at least 3 metres from existing trees or buildings. Mulch well to reduce competition from weeds. No artificial fertiliser should be given at planting time. It is a good idea to mix in some well rotted animal manure into the planting site at least six to eight weeks before planting out. Plant trees so that the strongest root and branch system faces south. Tree shelters may be required in more exposed gardens. Fertilisers should consist of a mix high in nitrogen and potassium, and low in phosphorus, NPK (13.2.14). Water frequently, especially during dry spring and early summer periods. Pruning should be aimed at establishing a strong, durable framework of main branches, avoiding the development of v-shaped crotches.

HARVEST

Flowering occurs in late winter and spring with nuts maturing in late autumn (however *M. integrifolia* may flower again in autumn and produce a second crop in spring). Before nuts begin to fall, it is a good idea to rake the ground clear. Harvest nuts within fourteen days of dropping so that they are not damaged by ground moisture. Regular harvests also minimise losses due to ravaging rodents and bandicoots. Mature nuts fall from the tree enclosed in a fleshy green husk. These husks should be removed within one or two days or kernel quality may be affected. It is advisable to dry nuts in a shallow even layer on wire racks indoors for two or three weeks before storage.

PROBLEMS

As the macadamia is native to Australia there are several insect pests. These include the macadamia leaf miner, macadamia twig girdler, flower eating caterpillar, macadamia nut borer and fruit spotting bug. Trees are tolerant of most diseases, with the exception of a trunk canker caused by *Phytophthora sp.*

CULINARY USE

Nuts may be eaten fresh or roasted. To roast nuts, place nutmeats in a shallow pan no more than two layers deep in an oven set at 120°-130°C. Lightly rub nuts in a little butter or coconut oil. When nuts are a dark cream colour they are removed and sprinkled with some fine salt.

MAMEY SAPOTE

(*Pouteria sapota*)

The mamey sapote, *Pouteria sapota*, is a favourite fruit in its native Central American homeland. It is extensively grown in the West Indies and from northern South America to Mexico, where the fruits are popular eaten fresh or made into tasty sherbets, milk shakes, ice-creams and preserves.

Mamey sapote grows to be a large tree (up to 20 metres or more) in its tropical homeland, however is more commonly seen as an orchard specimen to 10 metres. It is a handsome, open-growing evergreen with a thick trunk and heavy branches. Leaves are shiny, dark green, and are clustered towards the branch tips. The fruit is oval-shaped with a thick, rough brown skin. The reddish-orange or pink flesh is aromatic, almost fibre free and contains between one and four large brown seeds.

Other names for the fruit include **mamey**, **grosse sapote**, **mamey colorado**, **mammee**, **sapote grande**, and **zabotillo**. The mamey sapote is a member of the wonderful Sapotaceae family which includes such fruits as the abiu, *Pouteria caimito*, the caimito, *Chrysophyllum cainito*, and others.

MICROCLIMATE

Mamey sapote is a tropical tree which dislikes low temperatures. Young trees are frost tender and may die during very cold spells. Mature trees often shed their leaves after a light frost, and are severely damaged by heavy frosts. Trees will grow satisfactorily in a wide range of soils, providing they are free draining. Drought conditions often lead to temporary defoliation. A sunny, sheltered position is best.

VARIETIES

Seedling trees may take five to fifteen years to bear. Grafted selections should fruit in two or three years. Some varieties include: COPAN, MAGANA, MAYAPAN, PONTIN, TAZUMAL, CUBAN NO 1, CUBAN NO 2.

CULTURE

Plant out in a sunny, well drained position at least 5 metres from existing trees or buildings. Mulch well and provide individual climate shelters in more exposed gardens while permanent windbreak hedges are being established. Apply frequent light applications of poultry manure as fertiliser.

Little or no pruning is required. Water well during prolonged dry spells.

HARVEST

Fruit mostly mature in summer and early autumn months. It is rather difficult to determine just when fruits are ripe as the brown skin changes little in colour. You can wait until they fall naturally or try scratching the surface with your fingernail. If the flesh is still green underneath then leave them hanging. However, if it is a pink or orange colour then pick them and bring them indoors and they should continue to ripen and soften up in a few days.

PROBLEMS

Occasional pests include the sugarcane root borer, rhyparida beetles, scale insects, spider mites and termites. Anthracnose and rust fungus may damage leaves.

CULINARY USE

Mamey sapote is eaten fresh or made into ice-cream and milk-shakes, or blended with pineapple to make sherbets. Fruit flesh is rich and filling.

MAMMEA

(*Mammea americana*)

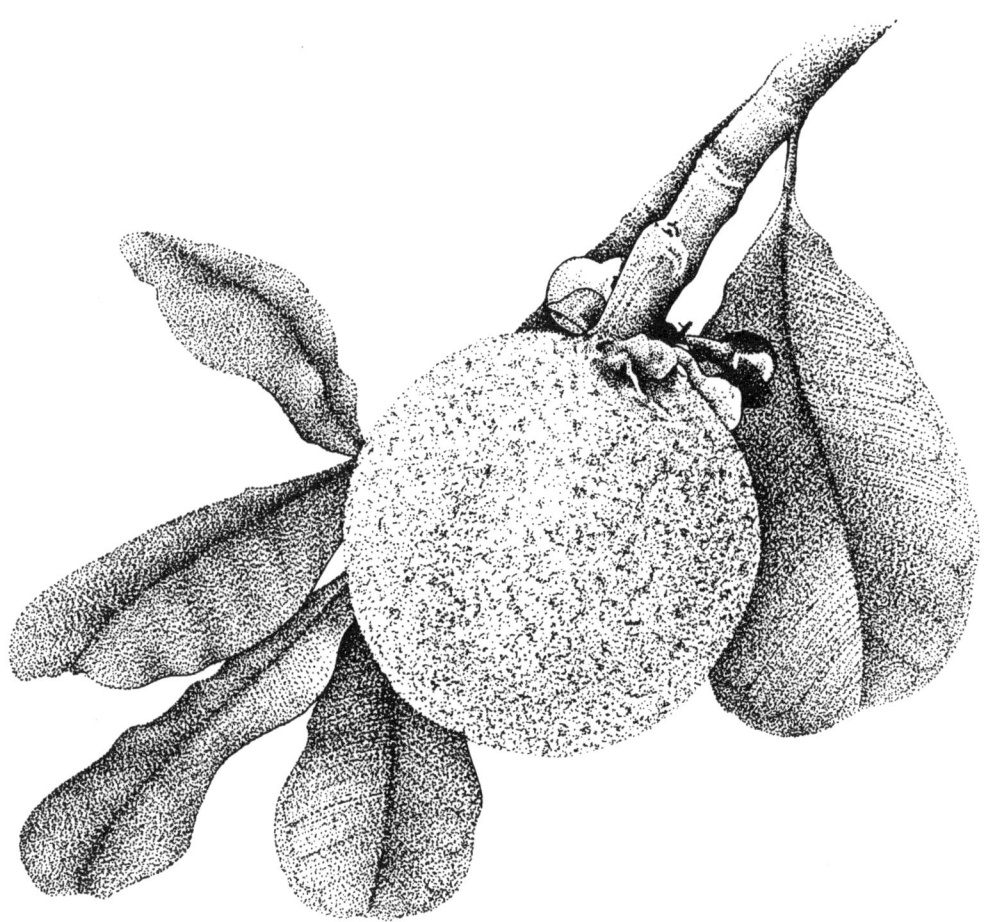

Mammea, *Mammea americana*, is a large, upright, evergreen fruit tree with thick, glossy, dark green leaves, and grows to a height of 15 to 25 metres under favourable conditions. The fruit is round and has a thick, rough, brown skin. The edible flesh inside is yellow-orange, firm textured, with a pleasant peach-apricot flavour. Each fruit has between one and four large, rough brown seeds.

Mammea is native to the West Indies and tropical America. It is eaten fresh as a dessert fruit, stewed, or made into jams and jellies. Other common names for the mammea include **mamee apple**, **mamey**, **mammey**, and the **apricot of San Domingo**. The tree belongs to the same family as the mangosteen, *Garcinia mangostana*, the madrono, *Rheedia madruno*, and the imbe, *Garcinia livingstonei*. The flowers are used to make a liqueur known as 'Eau de Creole' and the seeds and juice have been used to make an insecticide.

MICROCLIMATE

Mammea is a tropical tree preferring a hot-humid, frost free climate. Heavy frost usually kills young plants and severely damages mature trees. Trees grow best on a deep, free-draining, sandy loam with large amounts of organic matter. Young plants often prefer to be established in semi-shade. A warm, sheltered position is best.

VARIETIES

No named varieties currently available. Seedling trees may take up to six or seven years to fruit, and some may have no male flowers at all.

CULTURE

Plant out in warmer months at least 8 metres from existing trees or buildings. Water in well and mulch to help conserve soil moisture and retard weed growth. Young trees benefit from semi-shade from other trees or you may use climate shelters with shadecloth material. Soils should be kept moist. Fertilise with frequent, light applications of poultry manure. Little or no pruning is necessary.

HARVEST

Fruit are mainly harvested in summer months.

PROBLEMS

No serious pests or diseases affect production.

CULINARY USE

Mammea is eaten fresh on its own, or sliced and eaten with sugar and cream. Fruit is also popular stewed or made into jams and jellies. Immature fruit have a high pectin content and may be used to gel other fruits low in pectin and high in acid, such as the pineapple.

MANGO
(*Mangifera indica*)

The mango, *Mangifera indica*, is thought to be native to tropical and subtropical South-East Asia from the Indian-Burmese border region eastward to Indo-China. This fruit has been cultivated in India for over four thousand years, where it has played an important part in Hindu religious ceremony and folklore.

It is a large, evergreen tree varying in height from 4 to 20 metres, with a width of between 8 and 15 metres depending on growing conditions. Trees have a dense, dome-shaped canopy and look very colourful with each new growth flush.

Mangos come in many shapes and colours. Some varieties are oval-shaped, others round, oblong and some even banana-shaped. Fruit colour varies from green to yellow-orange to purple and red. Eating quality is mainly judged on freedom from fibre and lack of turpentine flavour. The fruit is highly esteemed in India, and indeed most other parts of the world. There are now more than five hundred known varieties of this ever popular fruit.

MICROCLIMATE

Trees perform well in a hot, sheltered, frost-free position on a wide variety of soils, providing drainage is satisfactory. Dry conditions in spring, during flowering and early fruit growth, are important to ensure adequate yields. When spring rains do occur production may be limited due to anthracnose and bacterial black spot disease which thrive in warm, moist conditions. Mangos tolerate high temperatures and are drought tolerant, however they are fairly cold sensitive. Young actively growing plants may be killed at a temperature of $-1°C$ to $-2°C$. Mature trees in dormant phase (winter months) usually escape damage, provided the temperature doesn't remain too low for too long.

VARIETIES

Mango varieties may be classified as monoembryonic or polyembryonic depending upon their origin. Most Indian varieties are monoembryonic, whilst the polyembryonic varieties are thought to have originated in Indo-China. The polyembryonic varieties, including the Bowen selections in Australia, may be propagated from seed. However, most of the improved varieties imported into this country are monoembryonic and require grafting or budding onto suitable rootstocks in order to propagate true to type. Some selections include: HADEN, ONO, GOLDEN TROPIC, FLORIGON, GLENN, EDWARD, IRWIN, ZILL, KENT, FASCELL, KEITT, NAM DOK MAI, BROOKS, SANTA ALEXANDRINA, VALENCIA PRIDE, TOMMY ATKINS and KENSINGTON PRIDE.

CULTURE

Plant out at least 6 metres from existing trees or buildings. Mulch plants and construct individual climate shelters in more exposed gardens. Avoid using fertilisers until plants are established and have experienced their first growth flush. Following this, plants should be given 0.1 kilogram ammonium sulphate in October, December, February and April for the first three years. In following years, trees should be given 0.5 kilogram NPK (10.10.10.) for every year of tree growth per annum applied in four equal applications. Avoid fertiliser applications for a short period just prior to flowering or yields may be decreased. Prune to develop a sturdy framework, with five or six main branches. Removing the terminal bud should result in a desired low spreading habit. Later pruning should involve the removal of weak, crowded or crossed branches and also any dead material present.

HARVEST

Trees commence bearing in three or four years with maximum production occurring when they are fifteen to twenty years old. Fruit mature from October through to April depending upon variety and location. Clip or cut fruit stems rather than pull or twist fruit from the tree. Ripeness is determined by cutting a segment of flesh and inspecting the colour of the pulp next to the seed. If yellow, then fruit should be ripe within seven days and harvesting should commence. Fruit can be cool stored satisfactorily at 10-12°C for two to three weeks.

PROBLEMS

Pests include monolepta beetles, fruit flies, scale insects, fruit spotting bugs, and seed weevils. Diseases include bacterial black spot and anthracnose fungus in wet areas. Spraying to control these diseases is essential.

CULINARY USE

Mangos are popular eaten fresh as a dessert with cream or ice-cream. They combine well with other fruits in salads, or they may be used in jams, chutneys, pies, cakes, puddings, mousses, sherbets and preserves.

MANGOSTEEN

(*Garcinia mangostana*)

The 'Queen of tropical fruits', *Garcinia mangostana* , is native to the tropical jungles of Malaysia, and also to Sumatra. It has been cultivated in South-East Asia for several centuries and many admirers describe it as Malaysia's most delicious fruit, being beyond superlatives. The tree is a broad-leafed, slow growing evergreen reaching a height of 12 to 20 metres. It makes a perfect specimen for the home garden with its dense foliage of dark, olive-green leaves and a lovely symmetrical shape that needs little or no pruning.

The round, purple-black skinned fruits contain four to eight white fleshed segments with one or two light brown seeds. The flesh has a sweet, melting, delicate and exquisite flavour. Mangosteen rind is very bitter and repels most insect pests. It is a source of a purple dye, and may be made into kampong toothpaste by charring, pulverising and mixing with a little camphor. Trees grow well along creek banks, and usually bear quite heavily. The mangosteen was named after a French botanist, Laurent Garcin, who first published a botanical description of this marvellous fruit. Mangosteen sorbet is a very popular and exclusive dessert dish.

MICROCLIMATE

The mangosteen prefers an equatorial climate with high temperatures and humidity and an even distribution of rainfall throughout the year. Sub-zero temperatures are usually not tolerated and there is some defoliation at temperatures below 5°C. Trees may die at temperatures less than 4°C, however some have been known to tolerate lower temperatures for short periods. Young trees up to four years should be given 30-50% shade, then gradually hardened into full sunshine. Soils should preferably be moisture-holding clay loams, mildly acidic with good quantities of organic matter. Temporary flooding and poor drainage is often tolerated.

VARIETIES

The flower of the mangosteen is 'perfect', and doesn't require the presence of pollen for fertilisation to take place. Fruit develops parthenocarpicly and seeds replicate the parent tree. Grafted trees have the advantage of fruiting in only one or two years, however trees may grow weakly and tend to bend over if not properly cared for. Seedling trees are often preferred, however they can take up to between ten and twenty years to fruit.

CULTURE

Plant out in the warmer months of the year at least 5 metres away from existing trees or buildings. Young trees should be planted into previously manured, moist soil, and should be watered in well. Provide 30-50% shade with shadecloth for about four years from planting out, then gradually introduce trees to full sunshine over a further twelve

month period. Individual climate shelters are a good idea in more exposed gardens as they offer short term wind protection while more permanent windbreaks are established. Little or no pruning is necessary, just remove any dead branches and inside growing shoots. Mulch regularly with compost and fertilise with a 20 millimetre dressing of poultry manure spread evenly under and just beyond the drip line every year after harvest. Water trees on a regular basis, however tend to neglect plants just prior to flowering.

HARVEST

There are two main flowering periods each year, with an October to December and a March to May crop. During maturation, fruit colour changes from green to red then purplish-black when ripe. Ripe fruit, suitable for immediate home use, may be left to fall naturally from the tree. This fruit should be eaten within two to four days. Fruit picked from the tree will last a little longer and may be left to ripen indoors over one to three weeks. Fruit may be cool stored at 10°C for up to eight weeks.

PROBLEMS

There are few pests due to a very bitter sap in the fruit skin and foliage. No serious diseases affect production.

CULINARY USE

Mangosteens are normally eaten fresh by themselves in order to fully appreciate their superb flavour. They are also used to make a tasty and refreshing sorbet. Cut the outer rind carefully so that the white inner flesh is not stained by the invading red, juicy latex.

MATISIA

(*Matisia cordata*)

Matisia, *Matisia cordata*, hails from the hot lowland forests of the lower Andes in Colombia, Venezuela, Ecuador and Peru. Matisia is also known as **South American sapote**, **quararibea**, **sapote colombiana**, **firolisto**, and **chupa-chupa**. The tree belongs to the family Bombacaceae which also includes the 'King of Malaysian fruit' the durian, *Durio zibethinus*.

Matisia is a medium sized evergreen that grows to 12 metres under favourable conditions. Trees have a tendency to spread with dense foliage consisting of some very lush looking, fan-shaped, deep green leaves. Fruits are round to oval with a greenish-brown leathery skin. The sweet, orange-yellow, fibrous pulp usually contains between two and five large seeds. Fruit taste varies from mango to rockmelon.

MICROCLIMATE

Matisia grows best in a hot-wet lowland tropical climate, however they should also perform satisfactorily in a warm subtropical climate. Trees are very cold sensitive and will either defoliate or succumb completely depending upon the severity of the frost. Shelter from cold wind is essential. Soils should preferably be deep, free draining, with a high organic matter content.

VARIETIES

There are no named varieties generally recommended at this stage. Seedling trees that produce fruit with less fibre are desirable.

CULTURE

Plant out in warmer months in a warm, sunny, sheltered position at least 5 metres from existing trees or buildings. Water in well to settle the soil around the roots, and mulch to retard weed growth. Individual climate shelters may be required in the more exposed gardens. Fertilise with frequent light dressings of poultry manure applied to the mulch surface. Water well during the normally dry spring and early summer months. Little or no pruning is necessary.

HARVEST

Matisia seedling trees take five to six years to fruit when grown under favourable conditions. Fruit are harvested in late summer and autumn. They ripen in one to three days and store well at 10°C.

PROBLEMS

Pests include the black beetle and the fruit spotting bug. No diseases seriously affect production.

CULINARY USE

Matisia fruits are eaten fresh or processed into refreshing drinks and juices.

MIRACLE FRUIT

(*Synsepalum dulcificum*)

From the tropical jungles of West Africa comes the amazing miracle fruit, *Synsepalum dulcificum*. It is a small, compact evergreen fruit tree or bush, with a dense cover of dark green leaves, and grows to a height of 2 to 5 metres. The plant bears a profusion of small, bright red, olive-shaped fruits with a sweet, white flesh, and a single smooth and shiny seed.

Miracle fruit acts on the sour receptors of the taste buds to turn all sour tasting foods sweet. Just a small amount of flesh sucked from each tiny fruit is sufficient, and the effect lasts for half an hour or more. The flavour of acid or sweet-acid foods is enhanced, but at the same time it does not affect the taste of sweet foods.

For a bit of good natured fun, try the miracle fruit out on your unsuspecting friends and visitors. Some of the sourest old lemons and limes from your orchard can be 'miraculously' transformed into the sweetest of treats!

Jakfruit. (D. Cilento/QTTC)

Top left
Red-angled tampoi. (D. Chandlee)

Above
Bambangan, dabai and engkala. (D. Chandlee)

Keranji papan. (D. Chandlee)

Pangkal. (D. Chandlee)

Hunting fallen durians.
(D. Chandlee)

Tutong, *Durio dulcis*.
(D. Chandlee)

Red-fleshed durian, *Durio graveolens*. (D. Chandlee)

Durian dealing on the waterways. (D. Chandlee)

MICROCLIMATE

The miracle fruit prefers a warm, humid climate. Light frosts usually result in some leaf defoliation, and may kill plants in more exposed situations. Partial shade and shelter from winds is beneficial to satisfactory growth. Soils should be acid and kept moist to help maintain constant humidity.

VARIETIES

Seedling trees should bear in two to four years depending on the growing site. There are no named varieties recommended at this stage, however there exists a small leaf type and a large leaf type.

CULTURE

Plant out in a warm, humid, sheltered spot, in filtered sunshine. Tub plantings are a good idea as you can obtain more control over your microclimate. Tubs are filled with soil mixed with a little peat moss to maintain acidity. Plants should be watered every three days, or as often as required to maintain soil dampness. They benefit by the addition of an organic fertiliser such as Nitrosol, seaweed, or fish emulsion. This should be applied in frequent light applications (at least every second week). Signs of over-fertilisation are the appearance of small, malformed, waxy leaves. Tub plantings give you the flexibility to move plants from positions in the garden to the verandah or the glass house, depending on the seasonal weather conditions.

HARVEST

The main harvesting period is in winter months, however some bushes will bear almost all year round with a short resting period of four to six weeks. From flowering to fruit is only three or four weeks, so several crops are produced in each season. Fruits will keep satisfactorily in the refrigerator, or may be de-seeded and frozen for future use.

PROBLEMS

No serious pests or diseases affect production. Scale insects may be an occasional pest.

CULINARY USE

Just a little of the flesh sucked from a single fruit is enough to sweeten the taste of your sourest foods. The flavour of strawberries, pineapples or grapefruit is delightfully enhanced and the effect lasts for thirty minutes or more.

PAPAYA

(*Carica papaya*)

This marvellous fruit is also known to most Australians as **papaw** or **paw paw**. In days past the Central American Indians found that they could tenderise their meat if they rubbed it with the juice of the papaya fruit, or wrapped it in the plant's leaves before cooking. They also found that they could consume copious quantities of food and not suffer from the discomforts of indigestion, providing they ate papaya for dessert.

The papaya, *Carica papaya*, is indigenous to the lowlands of Central America, probably from southern Mexico and Costa Rica. It is a small, single stemmed, herbaceous tree which grows vigorously to a height of 3 to 10 metres, and has a broad, shady crown of large, deeply lobed, green leaves clustered in spiral fashion at the top of the stems. Fruits are round to oval-shaped and have a smooth, thin, pale yellow to reddish-orange skin. The yellow-orange flesh within is firm textured and sweet tasting in the better varieties. Hundreds of small, shiny black seeds are found massed together inside the hollow centre of each fruit.

MICROCLIMATE

Papayas grow satisfactorily in a wide range of areas from the equatorial tropics to temperate latitudes. However, they must be grown in warm, sunny positions sheltered from wind. Exposure to light frosts or cold winds usually results in leaf damage to trees of all ages. Severe frosts will usually kill trees, with young ones being the most susceptible. Soils should be free-draining as the papaya is prone to root rots which thrive in waterlogged situations. A deep, well drained soil rich in organic matter is ideal.

VARIETIES

There are three main types of papaya plants. Female plants produce large flowers close to the trunk which after fertilisation develop into large, smooth, oval-shaped fruit. Male plants manufacture the pollen required for fertilisation of female plants. Bi-sexual plants are self pollinating and are suited to smaller gardens in the hotter climates. Some varieties include: RICHTER GOLD, HAWAIIAN SOLO (bi-sexual). Papayas may also be home propagated by washing the pulp from the seeds, and sowing them in a sandy soil mix.

CULTURE

Plant out in warmer months at least 1 metre from existing trees or buildings. There is no reliable method of determining sex of papaya plants until first flowering occurs, so at least four or five seedlings should be planted in the one planting space. From this group one plant is retained (not necessarily the strongest as these are often male plants). One male tree is sufficient to pollinate up to nine females. Mulch around each young tree and water well during the growing months. Animal manure is useful as fertiliser when added to the mulch. Trees over three or four years old are often cut back to 1 or 2 metres from ground level to make harvesting easier. This is done in spring months.

HARVEST

Papayas commence bearing eighteen or nineteen months or less after planting out. The main harvest period is from spring to early autumn. Be careful when picking fruit to avoid the white latex exudate which is a well-known skin irritant. Fruit is best harvested just before it is fully ripe and taken indoors to soften up fully at room temperature.

PROBLEMS

Major pests include the broad mite, the cucumber fly, red spider and fruit spotting bugs. Diseases include damping off, die back, powdery mildew, root rot, anthracnose and various fruit rots.

CULINARY USE

Papaya is eaten on its own or more popularly used in fruit salads with cream or ice-cream. It is also used to make sherbets, ices and milk shakes. In the West Indies unripe fruit is cooked and served as a vegetable. Jams, pickles and chutney are also popular.

PEPINO

(*Solanum muricatum*)

The pepino or **melon pear**, *Solanum muricatum*, is a member of the Solanaceae family, which also includes the cocona and the naranjilla. It is native to the temperate Andean regions of Colombia, Peru and Chile in South America. The fruit is a common sight in the market places there.

Pepino is a small, herbaceous, biennial fruiting plant or bush growing to a height of about 1 metre when trained on a trellis. The plant's growth habit is similar to that of a dwarf tomato. Pepino fruit are round to egg-shaped and have a rather attractive yellow skin which is commonly marked with numerous purple streaks or stripes. Better quality fruit is moderately sweet, refreshing and very juicy with a taste and aroma similar to rockmelon.

MICROCLIMATE

Pepino is suited to a warm temperate or subtropical climate. It is a hardy plant that likes a sunny or semi-shaded, sheltered, frost free position with fertile, free draining soils. Supplementary watering is necessary during hot, dry spells, especially for bearing plants.

VARIETIES

Some varieties include: SCHMIDT, EL CAMINO, GOLDEN SPLENDOUR, NARAGOLD, WAYFARER SPECIAL, KENDALL GOLD and SUMA.

CULTURE

Planting is carried out in spring months in your vegetable garden. A spacing of about 0.6 metres between bushes is recommended. It is a good idea to mix in some well-rotted poultry manure to the planting site several weeks in advance. Mulch well to suppress weed growth. Water young plants frequently, especially during dry periods. About six weeks from planting out, a standard 2 wire tomato trellis about 1.5 metres high may be constructed to help train and support bushes. Wire netting placed horizontally above the ground may also be used, so long as fruit is not allowed to touch the soil below.

HARVEST

Pepinos planted in early spring should fruit the following summer and autumn. In warmer areas bushes may fruit all year round. Ripe fruit should be handled with care as it bruises easily. Fruit can be kept satisfactorily at room temperature for one to two weeks or up to four weeks in the refrigerator.

PROBLEMS

Pests include weevils, caterpillars, mites and fruit fly. Diseases include blight, wilt and fruit rots.

CULINARY USE

Pepinos are refreshing when eaten fresh with a little sugar or lemon juice. The juice is good when mixed with more acid juices such as citrus.

PERSIMMON

(*Diospyros kaki*)

The persimmon, *Diospyros kaki*, from China, is a slow growing, deciduous fruit tree, growing to a height and spread of up to 5 metres or more. It is a highly ornamental specimen with its large, glossy green leaves which turn to lovely autumn shades before falling.

Fruits are rather similar in appearance to the tomato and are coloured bright yellow-orange to red depending upon variety. There are two main types of persimmon fruit, astringent and non-astringent. Astringent types cannot be eaten until fully ripe due to their high tannin content. Non astringent fruit can be eaten firm and crisp like apples. Ripe persimmons are sweet and aromatic, with a pleasant jelly-like consistency, and are delicious when eaten by themselves or mixed with pineapple.

MICROCLIMATE

The persimmon adapts well to a wide range of climates, from subtropical to warm temperate, and tolerates cold, frosty conditions during winter months. Non-astringent types need warmer conditions than astringent types to produce good quality fruit (average autumn temperatures 16-22°C). The tree has a low chilling requirement and can be grown satisfactorily in most subtropical areas. A well drained, sandy loam is preferred with at least 1 metre of free drainage, however trees do tolerate heavy soil conditions. A sunny position, sheltered from strong winds is recommended.

VARIETIES

Dwarf and semi-dwarf trees are precocious bearers and should fruit in two to three years. Non-astringent varieties include: FUYU (semi-dwarf), HANA FUYU (semi-dwarf), POMELO 152-7 (semi-dwarf), IZU (dwarf), MAEKAWA JIRO (dwarf) and ICHIKIKEI JIRO (very dwarf). Astringent varieties include: FLAT SEEDLESS (large), DAI DAI MARU, HACHIYA (medium) and TANENASHI (large). Non-astringent varieties may need pollinator trees to set fruit. These include: DAI DAI MARU, ZENGIMARU and GAILEY.

CULTURE

Plant out in a sunny, well drained spot at least 3 to 5 metres (dwarf varieties closer) from existing trees or buildings. Trees should be planted out during the warmer months of the year. Disturb the root system as little as possible and keep well watered making sure that the root system never dries out. Young trees may need staking or individual climate shelters in windy locations. Mulch regularly and apply 0.5 kilogram of complete fertiliser per year of age each year before budbreak in spring. If fruit drop or dehiscence occurs reduce nitrogen levels. Water well from fruit set through to harvest. Prune sparingly or crop yields will be reduced. Dwarf and semi-dwarf varieties should be trained to a central leader, with well spaced laterals (every 0.5 metre) to avoid fruit rub. Summer pruning of mature trees should thicken laterals and increase fruit size and colour.

HARVEST

Fruit are harvested from February to May. They should be clipped from the tree when the full skin colour has developed and the fruit is still firm. Astringent varieties should be left to soften up fully before eating. Fruit picked too soon lacks flavour and sweetness.

PROBLEMS

Pests include the fruit fly, fruit bats, fruit spotting bugs, soft brown scale, mealy bugs, thrips and birds. Diseases include crown gall, leaf spot, root and collar rots.

CULINARY USE

Persimmons are delicious eaten fresh, especially at breakfast time. The sweet pulp is also made into cakes, ice-cream, ices, pies, pureed sauces and dressings. A refreshing tea, high in vitamin C, is made from the leaves.

PUMMELO

(*Citrus maxima*)

Malaysia and Thailand are home to this giant sized citrus fruit. The pummelo, *Citrus maxima*, is prized as a dessert fruit in most South-East Asian countries. Some very good pummelos are grown in Thailand, sweet as an orange with an excellent flavour.

Pummelo is a bushy, spreading evergreen, growing to a height of up to 10 metres with shiny, dark green leaves. The tree and fruit is similar in appearance to the grapefruit, *C. paradisi*, the main difference being pummelo's larger leaves and extra large fruit.

Fruits are very large, round to pear-shaped, and have a thick, light green to yellow rind. The edible flesh is cream, pink, or red and is divided into ten to fifteen readily separable segments. In better varieties it is sweet and juicy, mildly acidic, but without the bitterness of grapefruit. Another common name used for this fruit is the **shaddock**, named after Captain Shaddock, the commander of an East India ship, who introduced the pummelo into Barbados in the seventeenth century.

MICROCLIMATE

Trees grow best in the humid, frost free, tropical lowlands, however some varieties perform satisfactorily in a subtropical climate. Soils should be free draining sandy loams of moderate acidity. A warm position sheltered from cold winds is preferable. Saline conditions are tolerated.

VARIETIES

Seedling trees are often variable and are not generally recommended. Some improved selections include: BOSWORTH PINK, STRATFORD, GRICE, SHATIEN, KHAO PHUANG and CHANDLER.

CULTURE

Plant out in warmer months at least 4 to 6 metres from existing trees or buildings. Mulch well to conserve soil moisture and discourage weed growth. Frequent light applications of a citrus fertiliser should be applied from fruit set through to harvest. Trees should be pruned to keep branches from touching one another and to remove any dead material. It may be necessary to prop up heavy-bearing branches with poles to prevent limb breakage.

HARVEST

Pummelos should begin to bear in two or three years from planting out. Fruit is picked as soon as it ripens. In Thailand it is picked when it starts to lose its dark green colour and stored indoors for one or two months to improve its flavour and juiciness.

PROBLEMS

Same pests and diseases as for other citrus types. Scale insects, fruit fly, bronze orange bug. Collar and root rots.

CULINARY USE

Sweeter varieties are eaten fresh. The fruit is squeezed into a refreshing juice, and also makes a good preserve. In Malaysia the rind is preserved to make peel candy.

PURPLE PASSIONFRUIT

(*Passiflora edulis*)

The purple passionfruit, *Passiflora edulis*, from southern Brazil, is an attractive, evergreen woody vine with long stems and spiralling tendrils which help support its rampant climbing habit. It is an ideal home garden specimen with deeply lobed, dark green leaves and large purple-white flowers, providing quick cover for those unsightly paling fences and bare brick walls.

Fruits are slightly oval to round, and when ripe have tough, brittle, dark purple skins. The juicy yellow flesh inside is pleasantly aromatic, rather pulpy, somewhat acid, and contains many small, black, edible seeds.

Passionfruit pulp is a good source of vitamin A and C, riboflavin and niacin. Some other tasty passionfruit species include the yellow passionfruit, *P. edulis f. flavicarpa*, the granadilla, *P. quadrangularis*, and the banana passionfruit, *P. antioquiensis*.

MICROCLIMATE

The purple passionfruit grows best in a sunny position sheltered from cold winds, and free from frosts. Vines grow well at higher altitudes in tropical climates, and can be grown in well protected locations in cooler temperate areas. Growth is satisfactory on a wide range of soils, however they should be free draining as waterlogging leads to root rot diseases.

VARIETIES

Some varieties include: LACEY, PURPLE CHAMPION, NELLIE KELLY.

CULTURE

It is a good idea to dig in large quantities of animal manure to the soil in the planting site two or three months in advance. If soils are shallow and poorly drained it may be necessary to mound. Plant out in spring months adjacent to boundary fences, walls, and trellises, or wherever there is likely to be adequate support for the vigorous growing vines. Water in well to settle the soil around the young roots, apply some mulch, and shade from intense sunlight for the first week or so. Train the leading shoot vertically by pinching off any side shoots until it reaches a horizontal support. The growing tip is then removed and two side shoots are trained horizontally in either direction. Little or no pruning is required other than to remove excessive growth touching the ground below. Fertilise regularly with light dressings of poultry manure to the mulch. Water well during the growing months.

HARVEST

Vines should start bearing within one or two years from planting out. The main harvest season is in summer months, with the possibility of a winter crop as well, in warmer areas. Ripe fruit develops a dark purple colour and falls naturally from the vine. It can be picked just prior to this when it comes away easily when touched. Excess fruit can be frozen for future use.

PROBLEMS

Main problems due to PWV virus, collar rot, septoria spot, brown spot, rootknot nematode and fusarium wilt. Insect pests include fruit fly, citrus mealy bug, vine mites and bugs, and scale. Excessive soil nitrogen may lead to luxuriant vegetative growth at the expense of cropping.

CULINARY USE

Fresh fruits are cut in half and the succulent pulp is eaten with a spoon, mixed with other fruit in a salad, or used in drinks. The beauty of the passionfruit is that just a little pulp goes a long way. The fruit is also used to make cake icing, candy, ice-cream, jelly, mousses, sauces, sherbets, syrups, pies and pavlovas.

RAMBUTAN

(*Nephelium lappaceum*)

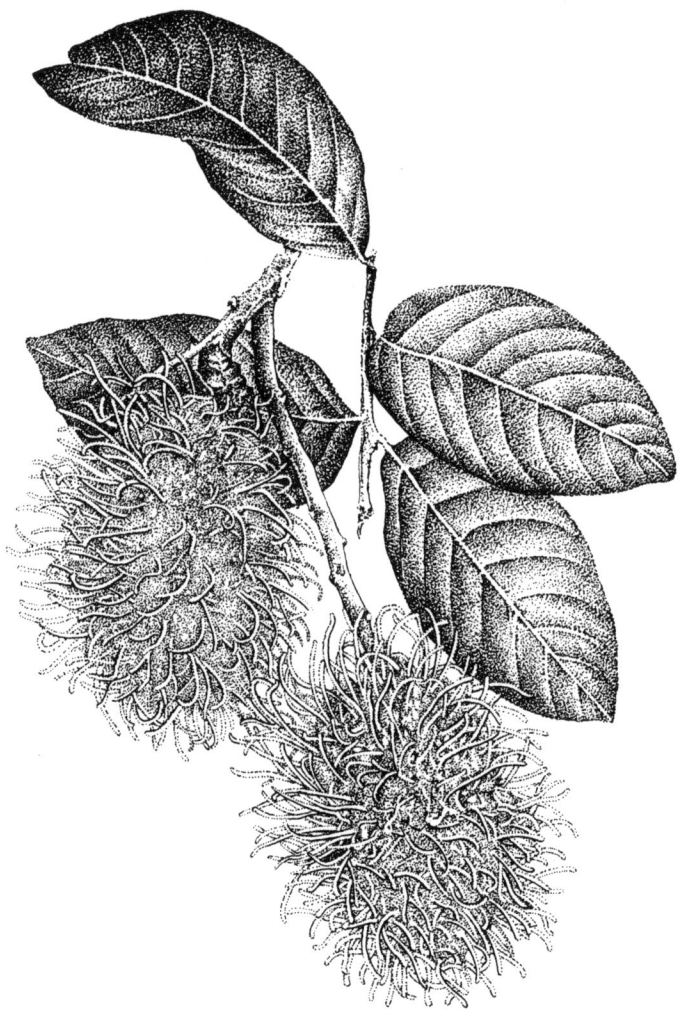

The rambutan, *Nephelium lappaceum*, or **hairy lychee**, is a medium-sized, bushy evergreen that is widely cultivated in lowland tropical South-East Asia. The tree is native to Malaysia and Sumatra and is found growing wild in the tropical jungles there. It makes a fine specimen for the home garden, particularly attractive when it is flowering and fruiting. Seedling trees grow to a height of 20 to 25 metres, however improved cultivars are smaller and more compact and rarely exceed 12 metres.

The common name, rambutan, is derived from the Malay word 'rambut' (meaning hair of the head), and refers to the thick covering of soft, red, pink or yellow spines on each fruit. Inside the skin is the sweet, white, translucent flesh which has a mild subacid flavour. It is very refreshing and has good quantities of vitamin C. There is a single, fibrous seed which is sometimes roasted.

Other names used for the rambutan include **ngo-phruan**, **litchi cheval**, and **rambutao**.

MICROCLIMATE

Rambutans prefer a hot-wet equatorial climate, with an even distribution of rainfall throughout the year. For best results a temperature range of 15°C minimum to 35°C maximum is ideal. Low temperature and frost is usually not tolerated at any stage of plant growth. Shelter from wind is recommended as it helps to create a more humid microclimate, as well as protecting trees from damage. Plants grow best in full sunshine once they are six months old. They are hardy and grow successfully in a wide range of soils as long as they don't suffer from waterlogging too frequently. Rich clay loams are often preferred, mildly acidic with a pH 5-6.5.

VARIETIES

Seedling rambutans may take up to five years to fruit. Improved cultivars are recommended and should take three to five years or less for the first crop. Some varieties include: CHOMPOO, RONGRIEN, RAPIAH, BINJAI, LEBAK BULUS, SILENGKENG, JIT LEE, MAHARLIKA, SEEMATJAAN, R3, R4, R7, R9, R37, R99, R134, R156, R161, R162, R163, R168 and R170. If you are having fruiting problems, then try planting a few different varieties together for cross pollination.

CULTURE

Plant out in the warmer months of the year at a distance of at least 4 metres from existing trees or buildings. Mulch trees well.

Construct individual climate shelters in more exposed gardens, while more permanent windbreaks are being formed. Fertilise trees annually with chicken manure spread evenly under the tree canopy and extending just beyond the drip line. Apply this after harvest. Water well all year except for a short period just prior to flowering. Prune fruit panicles back about 200 millimetres during or after harvest.

HARVEST

Trees flower from spring to summer, and fruit is picked from mid-summer to winter. Some trees may fruit all year round. Harvest fruit when they have coloured fully. Try one or two first and sample for sweetness. Clip fruit rather than twisting them off the tree. Ripe fruit lose their attractive colouring one or two days after picking, however they are still delicious and may be kept in the refrigerator for between one and two weeks.

PROBLEMS

Pests include rhyparida beetles, lepidopterous caterpillars, flying foxes and bush rats. Trunk canker may be a problem on young trees.

CULINARY USE

Rambutans are best eaten fresh, in order to appreciate their wonderful flavour. They are popular eaten alone, or mixed with other fruit in salads.

ROLLINIA

(Rollinia deliciosa)

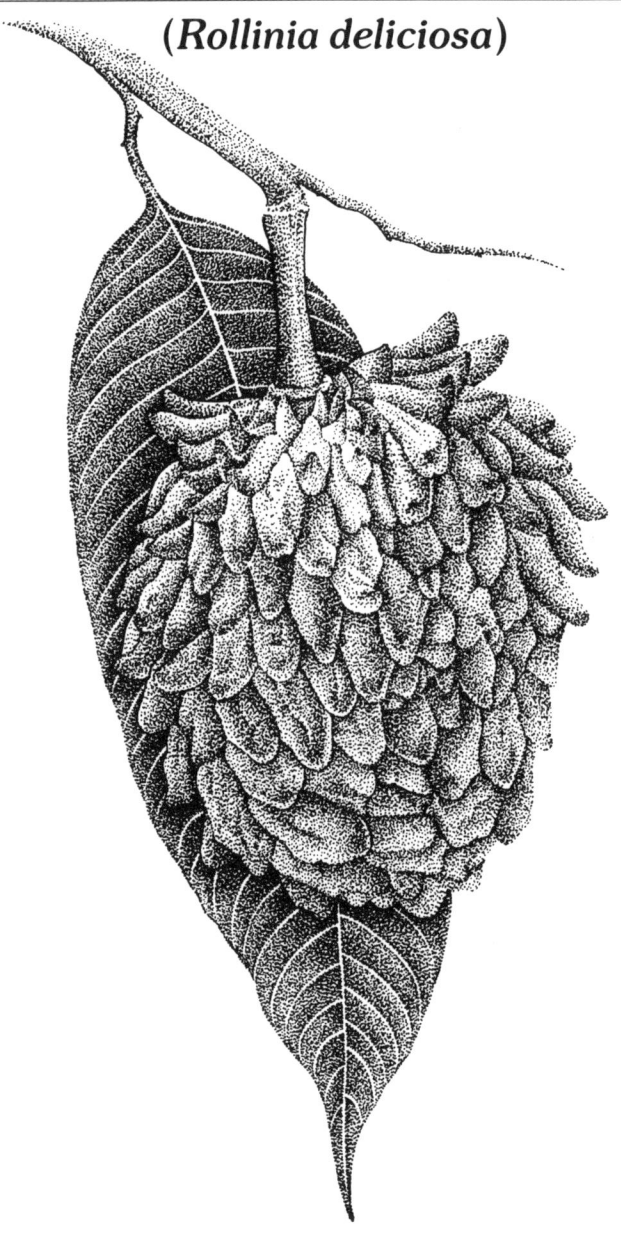

From the Amazon River regions in South America hails the rollinia or **Amazon custard apple**, *Rollinia deliciosa*. As the botanical name implies, this fruit is absolutely delicious, with a taste suggestive of a creamy lemon sherbet or a lemon meringue pie.

The yellow, oval-shaped fruit has a bumpy, spiny surface but a sweet, juicy, creamy-white flesh which just melts in your mouth. Each fruit may have up to eighty or ninety seeds.

Rollinia is a fast growing, semi-deciduous fruit tree occasionally found growing to 13 metres in its native Amazon habitat, however is more commonly seen as a small backyard tree with a height and spread of about 5 metres when mature. It is an attractive tree with its long, slender, feather-like branches and smooth, delicate, light green leaves.

MICROCLIMATE

Trees prefer a tropical or subtropical climate. Mature trees usually tolerate light frosts for short periods. Soils should be free draining to discourage root rots. Shelter from strong wind is beneficial as branches break and in wet weather the whole tree can be blown over. Trees grow well in full sun or partial shade.

VARIETIES

There are few named grafted varieties recommended at this stage. The sweeter fruiting types are preferred. Seedling trees should fruit in two or three years, or less, from planting out.

CULTURE

Plant out in warmer months, especially during periods of expected good rainfall, at least 4 metres from existing trees or buildings. Mulch lightly to discourage weed growth, however keep mulches away from the base of the tree to help prevent root rot. Trees have a high moisture requirement, but be careful not to overwater as they may succumb to root diseases. Fertilise with frequent light applications of poultry manure mixed into the mulch surface. Prune back long whippy branches. In more exposed gardens it is important to provide shelter from strong winds by constructing individual climate shelters and forming permanent windbreaks.

HARVEST

Rollinia fruit is mainly harvested from August to October months with some fruits being picked at odd times throughout the year. When fruit has just turned yellow, cut or clip it from the tree with secateurs leaving some stem attached. The best time to eat rollinia is when the fleshy spines have turned black. Fruit store best indoors in a cool spot at a temperature of 15-17°C.

PROBLEMS

The main disease problems occur when trees develop collar and root rots. Pests include fruit fly, mealy bugs, scale insects and leaf-eating caterpillars.

CULINARY USE

Rollinias are delicious eaten fresh by themselves, or blended with other fruit in salads.

SALAK

(*Salacca edulis*)

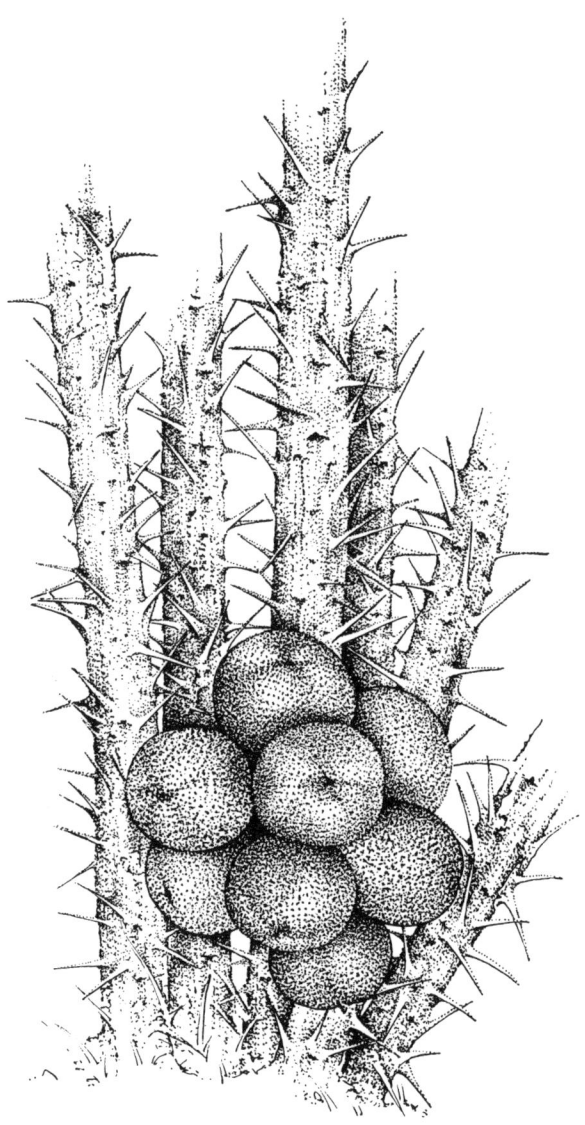

Salak, *Salacca edulis*, is a short growing palm (4 to 7 metres) from the rainforests of the Asiatic tropics, which displays an imposing armoury of thorns or spines and boasts a fruit with a surprisingly delicious flavour. Young palms have been known to sucker freely, whilst older plants have a tendency to sprawl along the ground, regenerating new trunks as they go.

Several scaly, light-brown skinned fruits are usually found clustered together at the base of each palm just above ground level. The edible flesh is a creamy yellow-white colour, crisp, with a pleasant sweet-acid flavour suggestive of pineapples. Each fruit possesses a single, brown, hard-shelled seed. Separate male and female plants are often necessary for fruiting, however the salak from Bali in Indonesia is self-fertile.

MICROCLIMATE

Salak is a rainforest specimen from the tropics preferring a hot-wet climate with daily temperatures ranging between 22°C and 32°C. Plants are cold sensitive and may be killed or severely damaged by heavy frost. A growing position in semi-shade is beneficial for young palms. Soils should be free-draining and rich in organic matter. Supplementary watering is recommended during long, dry spells. Windy locations should be avoided as salak has a short, fragile root system, and is prone to falling over under these conditions.

VARIETIES

The palms commence bearing in three years or less from planting out. The salak palms from Bali are often grown and are quite popular. Palms found growing in the swamps of Thailand bear elongated fruit on long stalks.

CULTURE

Plant out during the growing season, preferably during periods when good rainfall is expected. Young palms can be planted out at a spacing of 1.5 metres. Shade is beneficial (30-50%), using a shadecloth material or the protective cover of tall growing banana spp. such as Pisang Raja. Supplementary watering is necessary during long, dry periods. Seedling palms sucker enthusiastically for the first few seasons, and all but one of these suckers should be pruned off. As plants mature this tendency diminishes.

HARVEST

Fruits mature in late summer to autumn. Ripe fruit will keep for about seven days.

PROBLEMS

No serious pests or diseases affect production.

CULINARY USE

Salak fruits have a very appealing flavour when eaten fresh out of the hand. They are also canned in sugared brine, and green fruits are sometimes pickled.

SANTOL

(*Sandoricum koetjape*)

Santol, *Sandoricum koetjape*, is indigenous to Malaysia, Indonesia and the Philippines. In its native forest habitat santol is a lofty tree growing as high as 30 or 50 metres with a smooth, straight trunk. Home orchard specimens are usually much smaller and more bushy with a broad, shady canopy. The dark green leaves turn to some beautiful autumn shades of red or yellow when trees become semi-deciduous. Fruits are orange-yellow, round, and have thick, velvety skins. The soft, white, translucent flesh is juicy, thirst quenching, and in some varieties, has a rather peachy aroma. It clings very tenaciously to the seeds and makes for slow eating.

MICROCLIMATE

Santol is a hardy tree which can be grown successfully in tropical and warmer subtropical areas. They perform well in a hot-wet monsoonal climate with a regular dry season in winter and early spring. Light frosts may result only in some leaf fall, whereas heavier ones usually damage trees severely. Trees grow satisfactorily on a wide range of soils so long as they are free-draining. Shelter from strong winds is necessary to avoid limb breakage.

VARIETIES

There are two types of santol, red and yellow. Red santol leaves turn red at maturity and fruits are often sour flavoured. Yellow santol have leaves that colour yellow at maturity and fruits that generally have a sweeter pulp. Varieties include BANGKOK and MANILA.

CULTURE

Plant out in warmer months at least 6 metres from existing trees or buildings. Mulch well and apply frequent light applications of poultry manure as fertiliser. After plants are established watering is only necessary during extended dry periods. Young plants should be given individual climate shelters in cooler or more exposed areas to protect against frost and wind damage. Little or no pruning is required.

HARVEST

Trees start to fruit in about three to five years. Fruits mature from January to April. Picking normally commences once the first fruits begin to fall naturally from the tree. They will keep satisfactorily for at least seven days at room temperature.

PROBLEMS

Fruit spotting bugs and fruit sucking moths are an occasional pest. No diseases seriously affect production.

CULINARY USE

Santol fruit is eaten fresh or may be made into preserves and jams. It is also popular as candy.

SAPODILLA

(*Manilkara zapota*)

Sapodilla fruit is round to oval in shape with a russet brown, thin skin and a very sweet, aromatic, yellow-brown flesh, soft and melting, with a caramel or brown sugar taste. There are up to twelve shiny black seeds present.

The sapodilla, *Manilkara zapota*, is a hardy fruit tree originating in tropical Central America from southern Mexico to Venezuela. It is a handsome, upright, slow growing evergreen with thick, glossy-green leaves clustered at the tips of the branches. Seedling trees grow to a height of 12 to 30 metres, whereas grafted varieties are more compact and usually only reach a height of some 6 to 15 metres.

The sapodilla was discovered during the Spanish explorations where it was referred to by the Aztecs as 'chikl'. It was taken by the Spaniards to the Philippine Islands and is now found growing throughout most warm regions of the world including Florida, India, Malaysia, Thailand, Indonesia, Singapore and Australia. Other names for the sapodilla include **naseberry**, **nispero**, **chiku**, and **chiko**. Wood of the sapodilla tree is very tough and resilient and makes a very durable timber. In Mexico, Guatemala and Belize, sapodilla has been cultivated for a white latex (chicle), which was used as a base for chewing gum.

MICROCLIMATE

Trees grow well in a wide range of climatic conditions, from the wet tropics to dry, cool, subtropical areas. For best results they prefer a moist, hot climate similar to that found at medium to low elevations in tropical areas. Young trees are frost tender and may be killed at 0°C or below, whereas mature trees can withstand temperatures of about −3°C. Soils should be well drained, slightly alkaline, medium textured loams, however the sapodilla will tolerate a wide range of soil types from drier sands through to heavy clays with marginal drainage. Trees prefer full sunshine. They are fairly hardy and resistant to wind damage when mature.

VARIETIES

Seedling trees may take from five to eight years to bear, while grafted varieties usually only take two or three years from planting out. Some varieties include: KRASUEY, KAI HAHN, MARKOK, SAWO MANILA, PROLIFIC, BROWN SUGAR, H. C. TAN, TROPICAL, BKC 110, OVAL and MARTIN.

CULTURE

Plant out in warmer months at a distance of 3 or 4 metres from existing trees or buildings. Water in well to settle the soil around the roots and displace damaging air pockets. In more exposed areas young trees should be staked and tied or given individual climate shelters. Mulch well and water twice a week until plants are established, then only every one or two weeks depending upon soil type and weather conditions. Fertilise trees in March, August and December with 0.1 kilogram NPK (10.2.17) per tree for each year of age. It is also beneficial to give trees a light dressing of dolomite in August (0.5 kilogram per tree for each year of age) and urea in February (0.1 kilogram per tree for each year of age).

HARVEST

Sapodillas usually flower in summer months, however overlap sometimes occurs. Fruit is ready to pick in spring and the harvest extends through to late summer with a peak in October and November in more tropical areas. It is difficult to know just when to pick fruit as the only indication is a slight skin colour change to a lighter yellow-brown when the brown felt is scratched from the surface. Alternatively, you can wait until it falls from the tree. Pick the largest fruit. These should ripen in seven to ten days.

PROBLEMS

Sapodillas are generally free from major pests and diseases, however rhyparida beetles may damage new leaves and banana spotting bugs damage fruit, causing split lesions. Caterpillars may also be a nuisance.

CULINARY USE

Remove the fruit stalk and wash fruit to remove any sticky latex. Sapodillas may be eaten fresh out of the hand or made into sherbets, custards and ice-cream.

SWEETSOP

(*Annona squamosa*)

One of the most widely distributed members of the Annona genus is the sweetsop or **sugar apple**, *Annona squamosa*, native to Central and South America and now commonly grown at low to medium elevations throughout the tropical world. Sweetsop is a small, open growing, semi-deciduous tree or shrub growing to a height of 4 to 6 metres when mature. It is a hardy specimen, withstanding drought conditions quite well.

Fruits are heart-shaped with a rough, lumpy, yellow-green, red, or purple coloured skin. The sweet white flesh within is delicate and delicious, and usually conceals several small, shiny seeds. Sweetsop is a fine dessert fruit, eaten alone or with other fruits.

According to folk medicine a purgative tea is prepared from the tree roots and a tonic and mildly laxative tea from the leaves.

MICROCLIMATE

Sweetsop grows well in warmer areas of the subtropics, and at low to medium elevations in the tropical zone. Trees are drought hardy, however they usually drop some leaves and the fruits split, after extended dry periods. Young trees may be killed or severely damaged by light frosts, whereas mature specimens are a little more tolerant. Soils should be free draining but kept moist during the growing season. High humidity is often required for good fruit set. A warm to hot, semi-shaded position is preferred.

VARIETIES

Seedling trees should bear in two to four years from planting out. Budded or grafted plants are less variable in growth habit and fruit characteristics, however there are few named varieties recommended at this stage.

CULTURE

Plant out during the growing season at a distance of at least 3 metres from existing trees or buildings. It is a good idea to prepare the planting site two to three months in advance by digging in large amounts of animal manure and compost. Mulch well around each new plant to help conserve soil moisture and retard weed growth. Stake trees in more exposed gardens while permanent windbreaks are being formed. Regular applications of manure, NPK fertiliser and zinc are recommended. Water well from budbreak in spring right through to harvest in autumn. Be careful not to overwater though, as trees are prone to root rots. Prune back long, whippy branches during the growing season each year, and also thin out any dense crowns to enhance internal bud development.

HARVEST

Fruit mature mainly in autumn months, although odd fruits are often available throughout the year if conditions are favourable. The best time to pick fruit is when there is a lightening of the skin colour between the bumps. Allow them to soften and ripen at room temperature before eating.

PROBLEMS

Occasional pests include mealy bugs and whitefly. Diseases include root and collar rots.

CULINARY USE

Sweetsop is popular eaten alone or blended with other fruit in salads. It is also very tasty in sherbets, ice-cream and juice drinks.

SOME ADDITIONAL EXOTIC TREE FRUIT

AKEE
(*Blighia sapida*)

The akee is a very popular tree in Jamaica, however its true homeland is West Africa where it grows naturally in the forestland. It was transported to the West Indies during the late eighteenth century and is now a common sight there. The tree is a rapid growing, medium sized evergreen that reaches a height of 7 to 10 metres or more under favourable conditions. After fertilisation, the small, greenish-white flowers develop into thick, leathery, red or yellow fruit pods. These split open at maturity to expose three black, shiny seeds surrounded by the firm, creamy white, edible flesh or aril. The flesh resembles scrambled eggs after it has been cooked, and is popular sauteed in butter and served with codfish. Akee on toast is also a favourite. Immature and over-ripe fruit is poisonous. Never use the fruit that has fallen from the tree. Care should also be taken to remove the pink material that connects the aril to the seeds as this also causes illness. Trees grow well in tropical and warm subtropical areas.

AMBARELLA
(*Spondias cytherea*)

The ambarella is a rapid growing, semi-deciduous fruit tree from Polynesia with a smooth, straight trunk and large pinnate leaves. It is a medium sized tree, growing to between 12 and 25 metres, or more, when mature. Fruits are oval or egg-shaped with a leathery, orange-yellow skin and a firm, juicy, acid-sweet flesh with a mango-pineapple or apple-like flavour. Each fruit has a large, spiny stone that clings tightly to the flesh. Ripe fruit is eaten fresh, or used to make juice drinks, fruit butter, sauces and jam. Unripe fruit is used in salads, pickles or relishes. Ambarella is a tropical specimen that grows best in a warm, frost-free climate, however it will grow in warmer subtropical areas as long as trees are given some protection from cold winds and frosty temperatures when young.

ARABICA COFFEE
(*Coffea arabica*)

The coffee plant is found growing wild as an understory tree in the upland forests of Ethiopia, just a few degrees north of the equator. It is a shrub or small tree growing 3 to 5 metres high with glossy, evergreen leaves, fragrant white flowers, and attractive red berries. Each berry takes between seven and nine months to mature and contains two coffee beans. The highest quality 'gourmet' coffee berries are harvested before they begin to ferment on the tree. They are then pulped to extract the bean, and the beans processed to remove skin coatings. The end result is the dried green bean that is sold for premium prices on the international coffee markets. Coffee plants are frost tender. The optimum temperature range is 16-23°C. In tropical areas plants benefit from shade, however in cooler southern areas full sun is best. A warm, sheltered position, with a free draining soil is preferred. Plants fruit in their third year. In Ethiopia the dried coffee berries are chewed by the local inhabitants.

ASIAN PEAR
(*Pyrus pyrifolia*)

The Asian pear is also commonly referred to as **nashifruit** ('nashi' meaning pear in Japanese), **Oriental pear**, **Chinese pear**, **water pear**, **sand pear**, **apple pear** and **salad pear**. Most of us here in Australia are more familiar with the European pear, *Pyrus communis*. The Asian pear is apple-sized and more rounded than the European types. Ripe fruit is firm, crisp and juicy with a pleasantly sweet taste. The trees are frost and drought hardy, and also withstand periods of waterlogging. They adapt well to most soil types. The fruits ripen on the tree and are ready to pick when their colour changes from green to greenish-yellow. They are popular as a fresh fruit dessert.

BAEL FRUIT
(*Aegle marmelos*)

Also known as **Bengal quince**, **Indian bael** and **bel fruit**. The tree is native to India and is grown in most of the temple gardens there. Closely related to the citrus, it belongs to the same family. It is a small (3 to 6 metres), thorny, deciduous tree with trifoliate leaves and fragrant flowers. The orange-sized fruit have a hard skin and a soft, aromatic, seedy pulp with a pleasant flavour that is eaten fresh or dried. Refreshing drinks and sherbets are made from the fruit. Bael fruit is believed to have a beneficial effect on the digestive system. Growing conditions are as for citrus.

BAKUPARI
(*Rheedia brasiliensis*)

The bakupari is a Brazilian relative of the mangosteen. In Rio de Janeiro, these orange-yellow fruits are popular with the locals, especially when made into jam. It is a handsome tree bearing fruit with tough, leathery skins. The pulp inside is translucent, white and sub-acid, somewhat similar to the mangosteen. Other names for the bakupari include the **pacura** and **bacupari**.

BELL FRUIT
(*Syzygium aqueum*)

Also commonly referred to as the **water apple**, **water cherry** and **watery rose-apple**, the bell fruit is indigenous to India where it has been cultivated for several centuries. The small, bell-shaped fruit have pink or white, almost translucent skins which bruise easily. The edible flesh is sweet, crisp, watery and rather tasteless. Fruits are popular in salads and make good thirst quenchers. The tree is small (5 to 8 metres), with a low spreading habit, and is suited to a tropical or warm subtropical climate.

BILIMBI
(*Averrhoa bilimbi*)

Bilimbi is a close relative of the carambola, *Averrhoa carambola*. It is native to Malaysia where it is commonly grown in the kampongs. Trees are small, slender and quite graceful specimens, growing to 7 to 10 metres. Foliage is similar to that of carambola except that bilimbi has a paler green leaf, and with more leaflets per leaf. Bilimbi bears prolific quantities of small, yellow, gherkin or cucumber-like fruits, slightly five-angled and with a juicy, acid flesh which is best used for making pickles, preserves and curries. Trees grow well in a warm, frost free position sheltered from cold winds.

BRAZIL CHERRY
(*Eugenia uniflora*)

Brazil cherry is an attractive shrub or small tree growing compactly to a height of 4 or 5 metres. It makes a good hedge plant, looking very ornamental with a plum-red, new growth flush and some brightly coloured, deeply ribbed, cherry-like fruits. Trees and fruit grow well in a warm, moist climate and tolerate light frost when mature. They should be given plenty of water during dry periods. Fruit colour varies from a bright cherry red to almost black when fully ripe. The juicy, aromatic pulp is eaten fresh or used to make jams, jellies, pies and drinks.

CAMU CAMU
(*Myrciaria paraensis*)

The camu camu is found growing naturally along the moist river banks in the Amazon Basin. It is a small (4 to 6 metre), slow growing tree resembling its close relative the jaboticaba, *M. cauliflora*. The main difference is camu camu's larger leaves and more upright foliage. Fruits of the camu camu are very high in vitamin C and are processed into popular pink coloured drinks and organic vitamin tablets. They are round with dark red skins and have a very acid flesh similar to the lime. The fruits form along the sides of the terminal branches. Trees prefer a warm, frost free, sheltered position with a moist, free draining soil.

CAPULIN
(*Prunus serotina var. salicifolia*)

Capulin is a wild cherry from the subtropical highlands of Central America and northern South America. Trees grow to 10 metres with thick, rough barked trunks. They bear clusters of sweet, juicy, thin skinned cherries similar to the cherries that most of us are more familiar with. The capulin doesn't have such a high chilling requirement though, and grows well in a warmer climate. Trees tolerate light frost when established. They often take four to five years to bear fruit.

CAROB
(*Ceratonia siliqua*)

The carob is also referred to as **St John's bread** as it is believed that the 'locusts' that John the Baptist consumed in the wilderness were pods of the carob tree. It is found growing wild in the drier regions of the Mediterranean where it is valued almost as much as a shade tree as for its very sweet fruit. The carob grows to be a large, spreading evergreen in its native habitat, however grown elsewhere it is usually only a small, shrubby tree. The fruit is a dark brown, flattened pod containing a sweet, chocolate tasting pulp and several bean-like seeds. The carob is used in a similar manner to cocoa, in drinks, puddings, cakes and ice-cream. It is a cold hardy tree that grows well with little care and attention.

CHINESE JUJUBE
(*Ziziphus jujuba*)

Originating in northern China where it has been cultivated for four thousand years or more, the Chinese jujube is a small, deciduous tree that grows to 9 metres under favourable conditions. It is also commonly referred to as the **Chinese date**. The tree is a very hardy specimen withstanding drought, high soil salinity, waterlogging, and extremes of both heat and cold. The branches grow in a zig-zag pattern and generally have two spines at the base of each leaf, although there are spineless cultivars. The small, egg-shaped fruits are reddish-brown in colour and have thin, tough skins. The sweet, white, crisp flesh is a good source of vitamin C. Fruits are used fresh, dried or candied.

CHINESE RAISIN TREE
(*Hovenia dulcis*)

Although this tree is native to China, it is widely grown in Japan, and is often referred to as the **Japanese raisin tree**. It makes a good ornamental shade tree and has some attractive autumn foliage. The Chinese raisin tree is a small, deciduous specimen which grows to a height of 6 to 9 metres with elegant, heart-shaped leaves and small, greenish-white flowers. The fruits resemble dead twigs and are sweet tasting. They are commonly chewed by people of Oriental descent. The tree is very cold hardy and grows well with little care and attention.

COCONA
(*Solanum hyporhodium*)

The cocona is a small (1 or 2 metre), perennial, spectacular fruiting bush from the Upper Amazon with large leaves, and a spreading or sprawling habit. The cocona bears prolific quantities of large red or yellow, oval to egg-shaped fruits with a thick edible rind, and an acid, pale-cream flesh with a taste reminiscent of peaches. Fruits are clustered close to the trunk and branches, and bushes are often weighed down with as much as 30 kilograms each season. Due to its high acid, low sugar content it is usually eaten as a preserve or marmalade with a delicious tart and spicy flavour. The cocona grows best in full sunshine, is fairly drought resistant, and tolerates high temperatures. Plants prefer a frost free climate.

COCONUT
(*Cocos nucifera*)

The coconut is the most widely cultivated, and probably the most useful, of all palms. Generally thought to be native to Melanesia, it has spread to many other areas of the coastal tropics, floating with the ocean currents, or transported by man. It has been said that there are as many uses for the coconut as there are days in every year. Some of the useful things they provide include food, drink, medicine, oil, fibre, timber, domestic utensils, fuel and thatch. The slender, tall growing coconut palm is a tropical coastal specimen and grows unbranched to a height of up to 25 metres or more. Unless you are very agile, have a trained pet monkey or a knife attached to a very long pole, harvesting could be a problem! Not so with the 'dwarf' coconut cultivars as they don't grow as tall, and they also have the advantage of bearing fruit at an earlier age.

DATE
(*Phoenix dactylifera*)

The date palm is thought to be native to the Persian Gulf region. It is an oasis species and plays an important role in the diet of the desert nomads. It has been cultivated for thousands of years and is now widespread throughout the dry tropics. Palms grow to a height of 25 to 30 metres or more with an extensive root system. They tolerate drought, periodic flooding, salinity, and alkalinity. Best production occurs when temperatures are high (mean maximum 35°C in summer) during active growth and when rainfall and humidity are low during the fruit ripening period. Young date palms are often killed by frost. Fruits are eaten fresh or dried, cooked or uncooked. They have a high sugar content and are very nutritious.

GREEN SAPOTE
(*Pouteria viride*)

The green sapote is believed to be indigenous to the highlands of Guatemala, and also to Honduras. It is a similar tree to mamey sapote, *Pouteria sapota*, however it has smaller leaves and produces roundish, green-yellow to brown-green fruit with a thin skin. The edible flesh within is reddish-brown, sweet, juicy and smooth textured, with one or two seeds. The green sapote tree grows well up to medium elevations in the tropics, and in the warm, frost-free subtropics. Young trees may be severely damaged or killed outright in heavy frost conditions, whereas mature specimens may be defoliated and suffer branch or twig damage.

ICACO
(*Chrysobalanus icaco*)

Other names for icaco include **cocoplum** and **guinda**. Trees are found growing wild in coastal areas of tropical and subtropical America. Icaco is a compact, evergreen bush or small tree with thick, rounded, glossy green leaves. It grows to about 4 to 5 metres and is quite ornamental. Fruits are oval to egg-shaped with pinkish-white, dark purple, or red skins depending on type. The edible flesh inside is soft and white and makes good jams and jellies. Fruits may also be stewed. Trees are known to withstand short periods of waterlogging. They are not very cold hardy and may be killed by heavy frosts. Try to find a warm growing position if you can.

ILAMA
(*Annona diversifolia*)

From the mountains and foothills of south-western Mexico, Guatemala, and El Salvador comes the ilama. It is a small, slender tree that grows to about 7 or 8 metres. The foliage is similar to that of its close relative the sweetsop, *Annona squamosa*, except that the leaves of ilama are usually larger and more rounded at the apex. The oval-shaped fruits have a rough, sculptured surface, typical of the Annona family. Skin colour varies from pale green to purple. Edible flesh is sweet and may be a white or pink colour. Trees grow well in a warm, sheltered position and usually survive light frosts when mature.

IMBE
(*Garcinia livingstonei*)

The imbe is a small, asymmetrical tree from Portuguese East Africa with stiff, grey-green foliage and grows to a height of 3 metres or more. The small, orange-yellow skinned fruits have a thin, juicy pulp with a pleasant acid flavour and a single large seed. The trees are often valued for their ornamental beauty rather than for their fruit quality. They grow satisfactorily in most soils and have some drought tolerance. The pulp and the juice of imbe stain clothing. There are separate male and female trees.

IMBU
(*Spondias tuberosa*)

The imbu is a low spreading tree found growing wild on the dry plains of north-eastern Brazil. It is generally regarded as one of the best fruits of the genus Spondias, and has a green-yellow, thick skin and a soft, melting flesh. The trees are prolific bearers. The tree roots are also known to be edible. The imbu prefers a warm, frost free climate, is drought hardy, and grows to a height of about 6 metres when mature.

INDIAN JUJUBE
(*Ziziphus mauritiana*)

Indian jujube is a small, thorny tree with scented greenish flowers found growing in the warmer parts of India where it is referred to as **ber**. It is closely related to the Chinese jujube however differs in that it has a cream coloured, downy underside to its leaves, and is less cold hardy. Trees grow to 8 to 10 metres and can withstand hot desert conditions, sub-zero temperatures, and poor soils. Conditions of high humidity are not desirable and may result in reduced yields. Fruits are round to oval-shaped with yellow-orange to brown skins. The flesh inside is crisp and white and tastes somewhat like an apple. It is eaten fresh, dried or candied and is also made into sauces, jellies, puddings, cakes and breads.

INGA
(*Inga spp.*)

Also commonly known as the **ice cream bean**, the inga includes several species of medium to large sized trees from Central and South America, and the West Indies. They are all members of the family Leguminosae and are useful for fixing nitrogen in the soil. *I. edulis* has been used as a shade tree in coffee plantations in the hotter growing areas of the world. Some inga species are sought after for their edible fruits. These trees produce bean pods up to 1 metre long which contain the sweet, juicy, white pulp that is so popular in their native homeland. Inga is a tropical rainforest tree that grows rapidly and thrives in a moist soil. Some light frost is usually tolerated.

JAMBOLAN
(*Syzygium cumini*)

The jambolan is known by several other names including **jaman**, **lambol**, **duhat**, **Java plum**, **Jambolan plum**, **Malabar plum**, **Portuguese plum** and even **Indian blackberry**. India, Burma and Ceylon are home to this tall, handsome evergreen that grows quickly to a height of 12 to 18 metres and is useful as a windbreak or shade tree. The leaves are glossy and leathery, oval-shaped and dark green in colour. The shiny, dark purplish-red, oval-shaped fruit are often lacking in flavour. Some types are sweet tasting, while others can be rather astringent. A juice from the fruit is used to make jellies. The jambolan also makes a good preserve and an excellent wine. It is a good idea to wait until the fruits drop naturally from the tree before harvesting. Avoid planting jambolan too close to concrete paths or drives as fallen fruit can leave a troublesome stain. Trees are found growing in wetter tropical and subtropical regions, and withstand light frosts when mature. They also tolerate strong winds.

JELLY PALM
(*Butia capitata*)

Other common names used for the jelly palm include the **butia** or **wine palm**. It makes a good home garden specimen with its graceful pale blue-green foliage. Palms grow to a height of about 6 metres when mature, and the spreading habit of the leaves give them a bushy appearance. The jelly palm is native to Brazil and Uruguay. It is a fairly hardy specimen tolerating drought, winds and sub-zero temperatures, and grows well with minimum care and attention. The fruits are small, round to oval in shape, and are pinkish-yellow or orange in colour. They are eaten fresh, pureed, or used to make jellies and wine. Leave fruit in bunches on the palm and they should ripen several at a time. If whole bunches are harvested they tend to ripen all at once.

KEPEL FRUIT
(Stelechocarpus burahol)

The kepel fruit or **kepple apple** is native to Indonesia. It is said that the fruit makes one's body smell like violets. As folklore has it, the sultans in Java and Sumatra were the only ones allowed to grow this fruit, and they encouraged the ladies of the harem to consume large quantities. The fruit are brown and the size of golf balls.

KITEMBILLA
(Dovyalis hebecarpa)

The kitembilla is native to Sri Lanka and is also commonly referred to as the **Ceylon gooseberry** or **dovyalis**. It is an evergreen shrub that grows to 5 to 7 metres and bears clusters of small, round, reddish-purple fruits with a velvety surface. They have a rather acid, purple pulp, full of small seeds, high in vitamin C, and with a flavour reminiscent of the English gooseberry. Trees grow well in a warm, subtropical climate. Fruits are occasionally eaten fresh or more often used in jams, jellies, preserves, juices and custards.

LOQUAT
(Eriobotrya japonica)

The loquat is a small (6 to 9 metres), subtropical tree from China, that is now a common sight in many gardens throughout the world. It is a well shaped specimen, very decorative with its golden yellow fruits and dark green foliage, it also makes a good shade tree. Loquat is a frost hardy tree, however it grows and fruits best in subtropical areas with warm summers and mild winters. It can also be grown in cool temperate climates and at medium elevations in the tropics. Trees prefer a moist, fertile soil and a sunny, sheltered position. Fruits are eaten fresh, mixed with other fruit in a salad, or made into delicately flavoured jellies.

MADRONO
(Rheedia madruno)

The madrono is a small, handsome, conical-shaped tree from tropical South America with glossy, green leaves and medium sized, yellow fruits. They resemble a bush lemon and have tough, leathery skins and a white, sweet-acid pulp. The tree has a lovely symmetrical shape and looks very ornamental in the home garden.

MALAY APPLE
(Syzygium malaccense)

The Malay apple comes from Malaysia and bears pear-shaped fruit with thin, shiny, pinkish or dark red skins. The whitish flesh inside may be dry or moist. It has a rose scented aroma and a pleasant taste. There is normally one large seed. The Malay apple is also known as **pomerac**, **jambu merah**, **ohia** and **Malay roseapple**. It is a handsome tree growing up to 20 metres and is often planted in the tropics as a windbreak or ornamental, as well as for its fruits. Trees grow best in a moist, fertile soil. They are not very frost hardy and prefer a warm growing position. Some very attractive crimson-pink, 'shaving brush' flowers add to the ornamental beauty of this tree. Fruits are eaten raw, used in preserves or to make wine.

MAMON

(*Annona reticulata*)

Also known as the **bullock's heart**, the mamon is generally considered to be the true custard apple. It comes from tropical America where it is often seen as a backyard tree. The mamon is a small, spreading tree that grows to 6 or 8 metres, and is semi-deciduous losing most of its leaves in winter months. The smooth, reddish-brown, heart-shaped fruits have a creamy white, somewhat grainy textured dry pulp which is used in custards, ices and milk shakes. Mamon is a tropical specimen, however trees do grow satisfactorily in subtropical climates. Young plants are frost tender and should be given adequate protection. Soils must be free draining as waterlogging is not tolerated.

MAMONCILLO

(*Melicoccus bijugatus*)

Mamoncillo is a lowland tropical tree from the Carribean region in tropical America. There are several other common names used for this tree including **Spanish lime**, **honeyberry**, **Jamaica bullace plum**, **genip**, **knepe**, and **mamon**. It is an attractive evergreen with light green to blue-green foliage and greyish bark on the trunk and main branches. Trees grow slowly to reach a height of about 10 to 12 metres. They bear dense clusters of round or oblong, smooth green fruits with thin, tough, leathery skins. Edible pulp is soft, juicy, translucent, and is coloured white, cream or orange, the taste is sweet to acid, depending on type. Mamoncillo is refreshing when eaten out of the hand, and is also made into drinks. Fruits are sold on the streets and in the market places in Central America. A hardy tree, growing well on most soil types and during dry periods. Young trees may be killed by

sub-zero temperatures, whereas mature specimens usually survive one to two degrees of frost for brief periods.

MAPRANG

(*Bouea macrophylla*)

Native to the Asiatic tropics, maprang is the common name for this fruit in Thailand. Other names used include **kundangan**, **gandaria**, and **setar**. The maprang is a member of the Anacardiaceae family and resembles its close relative, the mango. The tree is smaller than the mango though, and forms opposite leaves along the stem (the mango forms alternate leaves). Maprang fruit is egg-shaped with a thin, brittle, yellow-orange skin. The edible flesh is sweet and juicy with a single, large seed. Trees prefer a tropical or warm subtropical climate.

MARANG

(*Artocarpus odoratissimus*)

The marang or **tarap** is a tropical tree from Borneo. It is a close relative of the breadfruit and bears fruit similar in appearance, however smaller in size. It grows to be a medium-sized tree with large, deeply-lobed, dark green leaves. Fruits are oval to oblong in shape with a yellow-green rind that is covered in soft, fleshy spines. Edible pulp is sweet, juicy, fibreless and melting in texture, resembling white grapes, as it surrounds each seed. Fruits develop a strong aroma after they ripen. The spines on the rind should snap when the fruit is ripe. If they bend and exude latex then they are still unripe. The fruit must be picked from the tree as it does not fall. Marang is a tropical specimen that prefers a warm, humid climate, and may be damaged by temperatures below 5°C.

MATASANO
(*Casimiroa tetrameria*)

Matasano is a close relative of the casimiroa, *Casimiroa edulis*, and is native to the highlands of Mexico and Central America. It is also known as **woolly leaf white sapote**, **yellow sapote** and **Guatemalan sapote**. Matasano differs from casimiroa in that is is less cold hardy, has a velvety leaf and generally bears fruit with a stronger, somewhat spicier flavour. The fruit usually has a yellow skin and a yellow or orange flesh when ripe. It is eaten fresh out of the hand or is pleasant as a dessert fruit with cream or ice-cream. It is also bottled in syrup, made into rich milkshakes or baked in pies.

NARANJILLA
(*Solanum quitoense*)

Naranjilla, the **golden fruit of the Andes**, is a small (1 to 3 metres), attractive shrub found growing naturally at higher altitudes in the northern Andes in Ecuador and Colombia. The naranjilla grows rapidly and fruits early, however only has a productive lifespan of two or three years. Plants have large, hairy green leaves and pale violet flowers. The round, peach-sized fruits have thick, orange skins covered in fine, brittle hairs. Slicing the fruit in half reveals a green, jelly-like pulp, similar in appearance to the kiwifruit, and containing numerous small, edible seeds. The flesh is pleasantly aromatic, with a rather acid taste. The juice is used to make the 'sorbete', a refreshing green foamy drink that is popular in the fruit's homeland. The fruit is also used in pies, preserves, and as flavouring in ice-cream and sherbets. Naranjilla is suited to a warm, frost-free climate and a growing position sheltered from strong wind. It grows well in a fertile, well drained soil.

NATAL PLUM
(*Carissa grandiflora*)

The Natal plum, also commonly known as **carissa**, is native to South Africa. It is a hardy, evergreen shrub (between 2 and 5 metres) with glossy, deep green leaves and white, star-shaped flowers that have a pleasant gardinia-like fragrance. Due to an abundance of sharp spines or thorns, Natal plum makes a very effective and almost impenetrable border hedge. Fruits are large, red and plum-like. They have a soft, pinkish flesh that contains a white, milky latex. Fully ripe fruits are eaten fresh or made into tasty pies and tarts. Slightly unripe fruits are used in jellies. The bushes have some salt tolerance and grow best in well drained soils. Young plants are frost tender, however they become more cold tolerant with age.

NUTMEG
(*Myristica fragrans*)

Nutmeg is indigenous to the eastern islands of the Moluccas where it thrives in the rich, volcanic soil and the hot-wet climate. The nutmeg is a spreading tree with a short trunk which grows to a height of between 5 and 13 metres, sometimes more. It has long, spear-shaped, shiny green leaves that are aromatic when crushed. Male and female flowers are borne on separate trees, although there are some bisexual types. The smooth, round to heart-shaped, yellow skinned fruits split open when ripe to reveal a shiny, purple-brown seed surrounded by a red, lattice-like aril or seed coat. The nutmeg spice is obtained from the seed, and mace from the aril. The fleshy pulp is popular when made into jellies and sweetmeats. Trees grow best in the hot, humid tropics. They require a free draining soil, a growing position sheltered from strong wind, and have little or no frost tolerance. Young trees benefit from shade.

Top
Pedalai and isau (*right*). (D. Chandlee)

Above
Pedalai. (D. Chandlee)

Top
Casimiroa. (B. Scomazzon)

Above
Santol. (B. Scomazzon)

Dabai tree in fruit.
(D. Chandlee)

Ripe (*foreground*)
and unripe dabai.
(D. Chandlee)

Left
Tampoi. (D. Chandlee)

Above
Purple meritams. (D. Chandlee)

Above
Garcinia hombriana. (D. Chandlee)

Left
Kubal tusu, *Willoughbea sp.* (D. Chandlee)

PANAMA BERRY
(*Muntingia calabura*)

Other common names for the panama berry include the **capulin** and the **strawberry tree**. The tree is indigenous to tropical America and the West Indies. It grows rapidly to a height of 7 to 10 metres, with a shady canopy of grey-green leaves and a profusion of fragrant white flowers similar to those of the strawberry. Fruits are smooth, red berries, very sweet and juicy, with numerous small, edible seeds. The trees should begin to bear in only six to nine months from planting out. They prefer a warm, sunny, sheltered position, free from heavy frosts.

PEJIBAYE
(*Bactris gasipaes*)

The pejibaye, or **peach palm**, is a slender feather palm growing to a height of up to 20 metres, with clumps of stems that are usually armed with sharp, black spines. The fruits possess a starchy, orange flesh and are generally eaten after boiling in salt water for a few hours. The heart of palm is also edible. Pejibaye is native to the Amazon regions of South America where it is a staple food in the diet of the local population for several months of the year. Fruit is sold in the streets as fast food. Pejibaye can be grown in a tropical or warm subtropical climate. It grows well in heavy clay soils and has an extensive root system. Palms may take up to four to eight years to fruit, however they are very heavy bearers. Pejibaye is a decorative, fruitful palm for the home garden.

PITAYA
(*Hylocereus guatemalensis*)

The pitaya is a vigorous, climbing cactus from Central and South America. This fascinating plant is not only a useful fruit producer but also puts on quite an eye-catching show with a beautiful display of highly scented, bell-shaped, nocturnal flowers. An aerial root system gains support on trellises, rocks, tree stumps or whatever is provided. There are two main species. The red pitaya bears oval-shaped fruit with a covering of leafy scales and has a sweet, red flesh. Fruits of the yellow pitaya are smaller, covered with clusters of spines, and have a translucent, white flesh with a delicious flavour. All fruits have a spongy pulp containing an abundance of small, black seeds. Pitayas are hardy specimens that grow satisfactorily in tropical, subtropical and temperate climates to 0°C.

PITOMBA
(*Eugenia luschnathiana*)

From Brazil comes the pitomba, a small, handsome, evergreen fruit tree which grows to a height of 6 to 9 metres with a tendency to spread. The tree has a dense cover of narrow, glossy leaves, deep green above, with a silvery sheen below. It bears a profusion of eye-catching, golden yellow fruits that have a juicy, sweet-acid, rose-scented pulp. They taste somewhat like an apricot and are eaten fresh or make good jellies and jams. The tree is quite hardy, tolerating light frost when mature and growing on a wide range of soils. The pitomba is considered to be one of the best of the Eugenias.

POMEGRANATE
(*Punica granatum*)

The pomegranate comes from Iran. For many years it has been a symbol of prosperity and fertility. Plants were grown in the hanging gardens of Babylon, and the fruit was known to the Romans as the '**apples of Carthage**'. They are now widely grown in most parts of the tropics and subtropics where they are valued as much for their ornamental beauty as they are for their yellow or red, leathery skinned fruits. The pomegranate is a small tree or bush which grows to a height of 3 to 7 metres. It looks very attractive with its glossy green leaves and orange-red flowers. Plants have a suckering habit and make a good hedge when closely planted. Fruit flesh is full of tender, edible seeds that are easy to eat and have a rather nutty flavour. The flesh itself is juicy and sub-acid. The seeds are often made into an attractive garnish, and juice is squeezed from the flesh.

RAMBAI
(*Baccaurea motleyana*)

Rambai is a South-East Asian fruit native to Malaysia and Sumatra. The tree is quite commonly grown in the village gardens there. An evergreen tree, it occasionally grows to 20 metres and bears its fruit on long stalks from the twigs, main branches and sometimes the upper trunk. Rambai fruit is similar to that of duku and langsat and is refreshing eaten alone or in salads. The fruits are round to slightly oval, with a pale yellow-brown, thin skin and a soft, white, translucent flesh which varies from acid to sweet depending on type. Each fruit has several flat, brown seeds. Rambai is a lowland tropical specimen that grows best in a warm, frost free position, and fruits in only three years from planting out. Separate male and female trees are necessary as the rambai is dioecious.

RAMONTCHI
(*Flacourtia indica*)

The ramontchi is also referred to as **Governor's plum** and **serali**. Indigenous to tropical southern Asia, ramontchi is a large, densely foliated, evergreen shrub growing to about 5 metres and makes an ideal hedge plant with attractive glossy, deep green leaves. Some plants have sharp spines. Fruits are round to flattish berries, ranging from dark red to purple in colour, with a soft, plum-like flesh and several small seeds. Ramontchi is eaten fresh or made into jams and jellies. It is a hardy shrub withstanding dry conditions and light frosts when established. They like full sun, a free draining soil, and benefit from top pruning. There are separate male and female trees.

RED MOMBIN
(*Spondias purpurea*)

Known also as the **Spanish plum** and the **purple mombin**, the red mombin is native to tropical America. It is a small, semi-deciduous tree which grows to 8 metres, often with a distinct spreading habit. The red or yellow skinned fruits have a juicy flesh and a rough stone containing several cavities. One of the best ways of eating the red mombin is to toss the whole fruit in your mouth and suck on the stone. Fruits are also made into jellies, syrup, wines and jams. In areas of the Carribean the trees are often seen growing as live fence posts. The red mombin may be grown in the tropics or warm subtropics, and tolerates light frosts when mature. Trees grow well on a wide variety of soils providing they are well drained. They have some drought tolerance and grow best in full sunshine.

ROSEAPPLE
(*Syzygium jambos*)

The roseapple is a small (7-10 metres) evergreen fruit tree from South-East Asia. It has an attractive pink growth flush which changes to a glossy, dark green as the leaves mature. The large, white or yellowish-white, showy flowers also add to the tree's ornamental beauty. The small, round to oval-shaped fruits have a green or white skin that is tinged with yellow or pink. The flesh inside is aromatic and often has a dry, crisp, sweet rosewater taste. Fruits are eaten raw or used to make jellies and confections. The roseapple is a favourite Indonesian fruit. The tree is a hardy specimen, growing well in the warm humid tropics and in the drier, cooler subtropical areas. A well drained soil is preferred. The wood from the tree, which is hard and heavy, is used for fuel.

TAMARILLO
(*Cyphomandra betacea*)

The tamarillo, also known as the **tree tomato**, is native to the high Andean country in Peru and Brazil. It is a fast growing evergreen shrub with large, tender, heart-shaped leaves that are covered in soft hairs. It has an erect growth habit and usually only grows to 2 or 3 metres, making it an ideal plant for smaller home gardens. It bears red to yellow, egg-shaped fruit some eighteen months from planting out. The fruit is a good source of vitamin C and has a sweet, sub-acid flavour. It is used in salads, sorbets, jams, jellies, juices, pickles and chutney. It is a good idea to blanch fruit in boiling water for a few minutes, then to peel off the tough, bitter skin before using. The yellow fruit is usually sweeter than the red skinned types. The tamarillo prefers a warm, frost-free growing site sheltered from strong winds. Plants have a shallow root system and benefit from the addition of a surface mulch to help conserve soil moisture. Soils should be free draining.

TAMARIND
(*Tamarindus indica*)

The tamarind tree is a large, spreading, semi-evergreen tropical legume growing to a height of up to 20 metres with graceful weeping branches that almost touch the ground. The long, brown, bean-like pods enclose a sticky, brown, sweet or acid pulp containing up to ten seeds. The pulp is commonly used to make refreshing drinks and savoury dishes such as curries, chutneys and preserves. It is also eaten fresh straight from the pod, and boiled with sugar and gelatin to make toffee. Native medicines are made from various parts of the tree including the leaves, flowers, bark, and roots. Tamarind trees may be found growing in their natural state in the dry savannah areas of tropical Africa, where they are used for shade and fruit. They were thought to have been introduced into northern Australia by trepang (sea-slug) hunters some five hundred years ago. The trees tolerate most soils, as long as they are free draining. Young trees are frost tender, however they develop a little more cold tolerance as they mature.

TAUN or DAWA
(*Pometia pinnata*)

Also known as **Fiji longan** and is found growing naturally in Papua New Guinea, Fiji, Western Samoa and the Solomon Islands. In Papua New Guinea the tree is utilised commercially for its timber. It grows to be a large tree to 30 metres or more in the jungles, however as a home garden specimen it is more shapely and only reaches about 15 metres. It has light green leaves which are deep red in colour when young. Fruits are round to oval-shaped and have green, red, purple, or brown skins when ripe. Edible flesh is white, juicy, semi-transparent and has a pleasant flavour suggestive of longan, with maybe a hint of something a little stronger. Trees are tropical in their requirements and prefer a warm aspect. They grow in poorly drained positions as well as on drier slopes and ridges, and seem to prefer an alkaline soil. Trees begin to bear in three to four years from seed.

UVILLA

(*Pourouma cecropiaefolia*)

The uvilla is also commonly known as the **Amazon tree grape** and is native to the western Amazon regions in Brazil, Colombia, and Peru. Trees produce prolific quantities of round, purple, grape-like fruits, clustered together in large bunches or racemes. When grown on dry, rocky hillsides the fruit flesh may be sweeter and less juicy than the flesh of fruit grown on more fertile soils, which tend to lack flavour. Uvilla grows rapidly in a warm, humid, frost-free climate and bears fruit in two or three years from planting out. One male tree is required for every nine female trees. The fruits are harvested over a lengthy period from October to March. They are eaten fresh or made into an excellent wine. The trees grow to a height of about 8 metres at maturity.

WAMPI

(*Clausena lansium*)

The wampi, from south China, is a small (7 to 10 metre), slender, upright tree that is quite

frost hardy and should grow well in tropical and subtropical climates. The tree makes a fine ornamental specimen, requires little pruning and adapts to both acid and alkaline soils. Wampi produces clusters of pale yellow, grape-sized fruit with an aromatic, jelly-like pulp sometimes containing up to five or six small seeds. They are eaten fresh as a dessert fruit or made into pies and jellies. They are very popular in China and are considered a cooling fruit.

WAX JAMBU

(*Syzygium samarangense*)

The wax jambu is a rapid growing, high yielding, spreading evergreen fruit tree that originates from West Malaysia and the Andaman Islands and reaches a height of some 10 to 14 metres. Trees are sometimes pruned to develop a broad, shady crown of glossy, dark green, water repellent leaves. Thin pieces of pinkish coloured bark flake from the trunk in an irregular pattern. Round to pear-shaped fruit with thin, waxy, yellow, white or red-pink skins, hang down in clusters from the small, leafy twigs. The edible flesh inside is crisp, succulent, slightly aromatic, but somewhat tasteless. Fresh fruits are often eaten raw with some salt. Wax jambu is native to the wet-humid tropics, however trees may also be grown in the warm subtropics. Mature trees usually tolerate light frosts to $-2°C$, though a warm, sunny, frost free position sheltered from strong winds is preferable.

RECENT RARE FRUIT DISCOVERIES IN MALAYSIAN BORNEO

Malaysia boasts a considerable diversity of rare fruits and other tropical crop plants. Many wild relatives of the more commonly cultivated species are to be found in the virgin forests of Malaysia. However, extensive logging operations are rapidly destroying these natural botanical habitats. In the past, Borneo's forests were managed for sustained timber production. They operated on an eighty year cycle and used small machinery which caused minimal damage to the forest ecosystem. More recently, poor management practices such as repeated overcutting and clear-felling with heavy equipment have resulted in wholesale destruction of these once-great forests. Where shifting cultivation follows the inroads made by the logging industry, serious topsoil erosion results and the virgin forest never returns. Where regrowth (secondary forest) occurs, it is usually impoverished and many unique and invaluable species are lost. Plant hunters and seed collectors are hastily searching these wilderness areas in the hope of saving some of this valuable flora before the devastation is complete.

A recent journey by David Chandlee and Lauren Gartrell, two Australian rare fruit explorers, to Sarawak and Sabah in Malaysian Borneo has successfully located many lesser known fruits with promise and interesting new specimens. Although many of these fruits have remained virtually unknown outside of their natural habitat, the local village people have collected them wild from the forests for several centuries. Commercial interest is now being shown in a number of them and it may not be too long before they are grown and appreciated in many other areas throughout the tropical world.

DURIO spp.

(*Wild or forest durians*)

TUTONG (*Durio dulcis*)

The tutong is also known as the **red durian** or **lahong**. This large, noble tree is found growing wild throughout the forests of Borneo. In its natural habitat it grows straight and tall to a height of up to 40 metres, often with a heavily buttressed root system. Leaves are glossy green with a flaky, gold undersurface. Large, pink flowers form in clusters on the older tree branches. The tutong bears some beautiful, red globular fruits (15 to 23 centimetres in diameter), covered in long, thin, 'needle-like' spines. The pale yellow, edible flesh inside is thick, creamy and delicious. It is soft and sweet like a good caramel and has a flavour of pineapple cream without the acidity. Tutong is the sweetest of all *Durio* fruits. It is difficult to open and is usually chopped transversely. The fruit shell has a strong odour.

Tutong grows well in the wet tropics under similar conditions to other *Durio* species. It is a rainforest specimen that is found growing in clay loam soils in the lowlands and along ridges. Microclimate establishment under *Gliricidia* trees (legumes that provide some quick shade) is recommended.

RED-FLESHED DURIAN (*Durio graveolens*)

From the rainforest jungles of Borneo, Sumatra and Peninsular Malaysia comes the red-fleshed durian, also referred to as **merahan** or **tabelak**. The tree grows up to a height of 45 metres, often with large, buttressed roots some 3 metres high. Leaves are glossy green with an undersurface of brown and gold scales. Small, white flowers are seen clustered along the older tree branches. The globular-shaped fruit (10 to 15 centimetres in diameter) are decorated with some sharp, closely spaced spines that are light red to orange in colour. Fruit flesh is bright red. It is very thick and creamy in texture, with a lack of sweetness that gives it an agreeable savoury quality. It can be eaten fresh, or cooked with onions and seasonings and used as a side dish with rice and vegetables. The fruit shell contains very little odour. The red-fleshed durian does not drop its fruits from the tree when ripe (unlike other *Durio* species) and the fruit must be picked when it has attained full colour. There also exists a natural hybrid (*D. graveolens* x *D. zibethinus*) that bears fruit with a mixture of characteristics from both species. The tree is found growing naturally on clay or sandy loam soils from sea level to an altitude of 1300 metres. It is now becoming a domesticated species and can be seen growing to a height of approximately 10 metres under cultivation in Sabah. Flowering occurs in about six years from planting out.

LAI or NYEKAK (*Durio kutejensis*)

The lai or nyekak comes from the forests of Borneo and is commonly cultivated in Kalimantan. It grows to be a small to medium-sized tree (up to 18 metres) in the Mixed Dipterocarp Forest, a very extensive and complex lowland forest which grows on mainly dry soils, the Dipterocarps being the leading tree family in Borneo. Leaves are glossy green with an undersurface of pale gold. Some large, red flowers (11 centimetres in diameter) make *D. kutejensis* a rather ornamental specimen. The tree is also referred to as **ukak** or **dalit**.

The small, round to ovoid, yellow fruits are covered in short, soft, pentagonal spines. There is a collar of whiskery hairs where the peduncle (stem) joins the fruit. The thick textured, orange to reddish-orange flesh is drier and firmer than the isu (*Durio sp.*). The flavour is distinctive; sweetish, but not as sweet or as strongly flavoured as the durian, *D. zibethinus*. It has a tangy aftertaste. A description taken from David and Lauren's travel journal reads: 'Sweet, isu flavour, with a touch of tutong. Texture dry until very ripe. Not sticky like isu'.

The lai or nyekak is a highly variable species. There are many varieties which differ significantly in fruit size and percentage of flesh. Although there are four named varieties in Kalimantan, it may be years before the others are sorted out. With good cultural practice in clay or sandy soil, trees should begin to fruit when they reach a height of 5 metres (about five years from planting out).

BELUDU (*Durio oxleyanus*)

The beludu is found growing wild in the rainforests of Borneo, Sumatra and Peninsular Malaysia. It is also cultivated in a small way by the local village people.

Another name used for the beludu is **kerantongan**. In its native forest habitat the tree grows tall and straight to a maximum height of 40 metres. Leaves are green and have a downy undersurface of blue-green with golden veins. The flowers are white. The round, greyish-green fruits (15 to 20 centimetres in diameter) are covered in stiff, pyramidal spines, slightly curved and up to 4 centimetres in length. The edible flesh inside is a very pale yellow colour and surrounds the reddish-brown seeds in each of the four locules. It has a rich, delicious flavour similar to that of the durian, *D. zibethinus*. The fruit does not have a strong odour.

The tree is normally found growing in moist locations, but exists on drier soils. It grows to an altitude of 1000 metres on Mt Kinabulu.

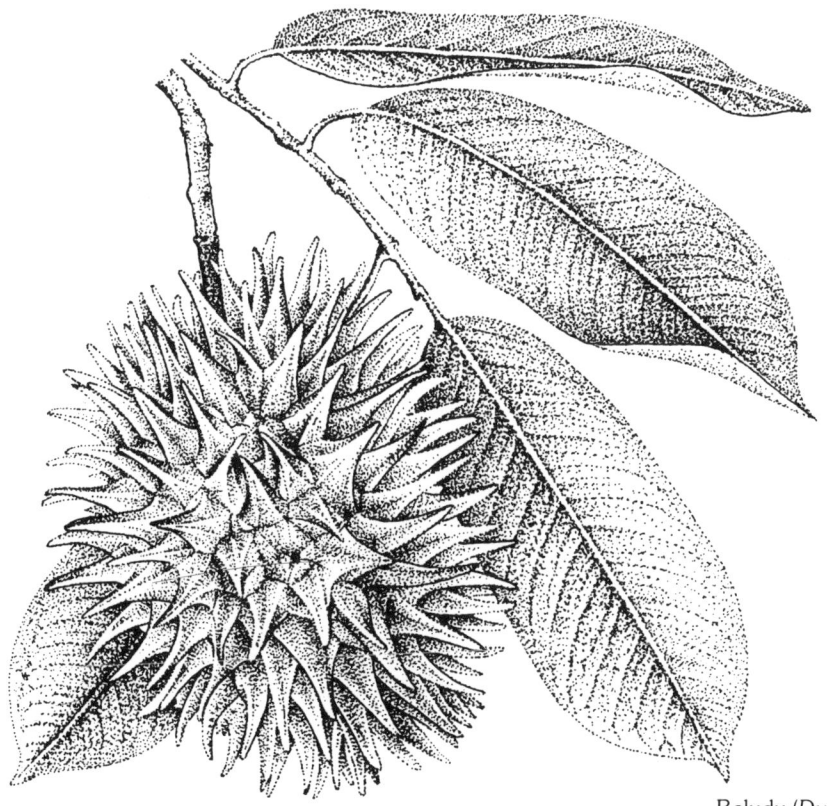

Beludu (*Durio oxleyanus*)

ISU, ISU RAMIN, ISU KUNING
(Durio sp.)

At this point in time botanists cannot decide just what the species is, so it remains a challenge for the future. The trees grow to a large size in the rainforests of Sarawak and, possibly, other parts of Borneo. Their leaves are glossy green with an undersurface of prominent, deep gold scales. There are at least three varieties of this species including isu, isu ramin, and isu kuning.

Isu fruits are small (13 centimetres diameter) and round with a green shell that often ripens to yellow. They are covered in long, sharp, pentagonal spines. The pale yellow flesh surrounds between one and three mid-brown seeds in each of the five locules. The flesh is a little thicker in texture than that of durian, *D. zibethinus*, and somewhat dry, like peanut butter. It has a good, rich flavour similar to that of *D. zibethinus*, but less sweet and has a slightly tangy aftertaste like most *Durio* species. The fruit shell doesn't smell strongly and is easy to open.

Isu ramin is similar in external appearance to isu and has a deep yellow flesh with a rich flavour. It is slightly sweeter than isu, with a thicker texture and drier flesh, like a fruity peanut butter. It tastes a little nutty too. The fruit shell smells strongly.

Isu kuning is similar to isu and isu ramin, with a combination of characteristics (such as sweetness and flavour) from both fruits. It has a mid-yellow flesh that is soft, thick and delicious.

ARTOCARPUS spp.

PEDALAI *(Artocarpus sericicarpus)*

The pedalai comes from northern Borneo, the Philippines, Sulawesi and the Moluccas. It is a very rare specimen in the Mixed Dipterocarp Forest of Borneo where it grows naturally up to a height of 40 metres. It is occasionally seen growing on the steep, clay hillsides in the inland regions. The foliage consists of very large (40 by 25 centimetres)

Pedalai (*Artocarpus sericicarpus*)

handsome, dark green leaves, spectacularly digitately lobed when young but entire when mature.

The pedalai bears some very beautiful, eye-catching, bright orange skinned fruits, globular in shape and about 15 centimetres in diameter. Small, soft protrusions on the fruit surface sprout curly, yellow hairs like a giant rambutan. Pedalai is similar to the marang inside, but has a superior flavour, firmer flesh and slightly larger segments. The sweet, creamy-white flesh is easy to eat and like the marang the segments cling to the central core when the skin is removed. As with all the *Artocarpus* species the seeds are edible and those of the pedalai are considered to be some of the tastiest. They may be boiled, roasted or fried. Fruit odour is not as strong as it is with the fruit of the marang, *Artocarpus odoratissimus*.

The tree grows rapidly in fertile, alluvial soils. Partial shade is beneficial during the early years of growth.

CHEMPEDAK (*Artocarpus integer*)

Other names used for this increasingly popular fruit include **temedak** or **nakan**. The chempedak occurs naturally throughout Borneo, Sulawesi, part of West Irian, Sumatra, and Malaysia to an altitude of 1400 metres. The tree is restricted to areas with no dry season. The foliage is similar to that of the jakfruit, *Artocarpus heterophyllus*. Chempedak fruits are like small, cylindrical versions of the jakfruit, however the skins are almost smooth, with only small protrusions. Internal appearance is similar to that of the jak, however with smaller segments (2 to 3 centimetres) surrounding each seed. These segments are attached to the central inedible core.

There are two distinct chempedak types: one larger, pale yellow fleshed fruit (30 to 45 centimetres in length) and another smaller, with dark yellow flesh (25 centimetres in length). Both types have a very rich, sweet, juicy flavour and a creamy texture. The smaller fruits are generally sweeter and creamier. The chempedak has a richer flavour with less fibre and acidity than the jakfruit. Among the small chempedaks there is an orange-fleshed variety and a green-fleshed type. The orange-fleshed fruit is sweet and creamy with an agreeable flavour.

The chempedak is good chilled and the Malays make an 'ice' out of it. Immature fruits are used as vegetables before their hard seed coats form. They are tender and flavoursome when boiled in a soup with onions and seasonings.

The trees are found growing wild in the Mixed Dipterocarp Forest on soils that are mainly clay types (including some very heavy ones). They are also widely cultivated, mainly propagated from seed, however some named varieties exist in western Malaysia. The chempedak is a very prolific cropper and fruits usually go to waste if their season overlaps with that of the durian, *Durio zibethinus*. Being a tropical specimen, the tree is cold sensitive and should be given microclimate protection in cooler climates.

ENTAWAK (*Artocarpus anisophyllus*)

The entawak occurs throughout Borneo, southern Malaysia and southern Sumatra. It is occasionally seen in the Mixed Dipterocarp Forest and Upper Dipterocarp Forest to a maximum altitude of 650 metres. It grows to be a medium to large tree (to 24 metres) in the primary forests. Leaves are large and deeply lobed.

The roundish, brown-yellow fruits (11 by 8 centimetres) are covered in closely set spines like the marang, *Artocarpus odoratissimus*. Edible flesh is sweet and orange-red in colour. The fruits are popular with the locals and are sometimes cultivated. Other common names used for the entawak include **kelidang**, **tawak**, **bintau** and **bintawak**.

SELANKING (*Artocarpus nitidus*)

This fruit is also referred to as **butong**, **empatah** and **sinojoh**. The tree is very widespread throughout South-East Asia from east India and south China to the Philippines, Borneo and Sumatra. It is a large specimen which grows to a height of 30 metres in the forests and savannah woodlands, from sea level to an altitude of 1500 metres. Leaves are small and lobed.

The smooth, soft, orange-skinned fruits (6 by 4 centimetres) have a sweet-acid, orange flesh with a few small seeds. Several sub-species exist.

PUDAU (*Artocarpus kemando*)

The pudau is a rare specimen that is occasionally seen growing throughout the primary and secondary forests of Sarawak. It also exists in other areas of Borneo, South-East Sumatra and Peninsular Malaysia. It grows to be a large tree in the forests, smaller under orchard conditions. Leaves (15 by 8 centimetres) are smooth on top, rough underneath. The small, round fruits (4 centimetres in diameter) have a smooth, green skin similar to the chempedak. The thin, edible flesh has a mild flavour without sweetness. Latex tapped from the tree is edible in small quantities. It is said to taste like coconut milk and is used as a sauce. Pudau grows in the lowlands but has also been seen on alluvial clay soils in the highlands at an altitude of 1000 metres. Other common names include **pudu**, **puroh**, **selibut**, and **chempedak ayer** (meaning water).

PINGAN (*Artocarpus sarawakensis*)

The pingan is a rare tree from Sarawak. Its foliage consists of some large (up to 60 centimetres), deep green, undivided leaves that are rather ornamental and have a covering of golden hairs on the petiole (leaf stalk).

The orange-brown fruits (75 to 100 millimetres in diameter) are covered in short, stubby projections like the marang. The flesh inside is similar to that of the marang, but the segments are smaller. The flavour is good, like the pedalai.

TEKALONG (*Artocarpus elasticus*)

The tekalong is seen growing in Sumatra, Java, Peninsular Malaysia and Borneo to an altitude of 1400 metres. It grows to be a very large tree (up to 65 metres) in the forests but is smaller under orchard conditions. It is a rare specimen in the Mixed Dipterocarp Forest and mainly grows on clay soils.

Tekalong fruits (17 by 10 centimetres in size) are similar in appearance to the pedalai, however they have a denser covering of shorter, reddish hairs. Edible flesh is sweet and white. The seeds may also be eaten when roasted.

Tekalong tolerates a short, dry season. Other names for this fruit include **terap**, **togop**, **benda**, **ahbat**, **jerami**, **mendi** and **ho**.

NEPHELIUM spp.

MERITAM or PULASAN (*Nephelium mutabile*)

This very popular and widely cultivated fruit is native to Borneo and Peninsular Malaysia. It is a rare specimen that is found growing naturally in the Mixed Dipterocarp Forest to an altitude of about 1100 metres, mainly on alluvial soils. The tree is very similar in appearance to the rambutan, *Nephelium lappaceum*, and usually grows to a maximum height of 12 metres under orchard conditions. Given favourable growing conditions it fruits in about five years on a 5 to 6 metre high tree.

The fruit are covered in whorls of small protruberances. They are larger, softer, juicier and more acidic than those of the rambutan. At the same time they remain quite sweet and refreshing. They are more flavoursome than lychees and are somewhat reminiscent of sweet, purple grapes. The

Meritam or Pulasan (*Nephelium mutabile*)

fruits can be eaten very green (four to six weeks before maturity) and still possess an excellent flavour, although they are a little more acidic than when fully ripe.

There are several varieties of meritam, with at least four found in Borneo. These include green, yellow, dark red and purple types. There is also a variety from West Malaysia that bears large sized fruit.

The culture and microclimate are similar to that required for rambutan. Establishment of plants under *Gliricidia* trees, or other suitable shelter is recommended. The meritam appears to bear fruit that is true to type (or nearly so) from seed. The **presen** is another species of *Nephelium* that is similar to the meritam, but with different foliage; it also bears good fruit.

LAIT (*Nephelium maingayi*)

The lait is native to Sarawak and Sabah. It is occasionally found growing in the Mixed Swamp Forest throughout Sarawak. It is a medium-sized tree, smaller than rambutan, with small, rounded, mid-green leaves arranged in pinnate fashion. This small (35 by 25 millimetres), juicy, rambutan-like fruit has a thin, red, grooved shell. The sweet, translucent flesh inside is slightly acidic with a delicious hint of mint flavour.

Little is known of the cultural requirements of this tree, but it is thought that it may tolerate some waterlogging.

SIBAU (*Nephelium sp.*)

Sibau comes from Sarawak where it is also referred to as **rupah**. Tree description is unknown but it is undoubtedly similar to the rambutan tree. The leaves are large (22 by 7 centimetres), dark green and elliptic with very long 'drip-tips'.

The fruit is like a small rambutan, good flavoured and sweet. Varieties include SIBAU RARAS and SIBAU MANOK. Sibau raras is unique among the *Nephelium* species. It is a dwarf variety that fruits when 1.2 metres high. It is very rare and grows in podzolic clays. Little is known about sibau manok.

PANGKAL (*Nephelium sp.*)

From southern Sarawak comes the pangkal, a small (10 metre) tree with large (15 by 8 centimetres), dark green, pinnate, obovate leaves. They have depressed veins and a downy undersurface. Tree bark is smooth. Pangkal is a very attractive specimen, especially when in fruit. The round, bright red fruits (50 millimetres in diameter) are often sour, however some varieties are sweet when fully ripe and taste like meritams. They are covered in short, stiff spines and the fruit shell is thick (5 millimetres). They are borne in beautiful, loose bunches at the ends of the branches. Edible flesh is usually thin and is melting in texture. Varietal selection will be necessary, as with all the minor Sapindaceae. Cultivate as for rambutan.

TITIDAN (*Nephelium sp.*)

The titidan is found growing in Sarawak. Tree description is unknown but it is undoubtedly a rambutan-like tree, with leaves that are larger and thicker. The small (25 millimetres long), oval-shaped fruits have a smooth, thin skin which is dark purple when ripe. The sweet, edible flesh is slightly acidic (most like that of the rambutan of the wild *Nepheliums*).

The common name, titidan, is derived from the tree's fruiting habit — the fruit grows in clusters along the branches, which is unusual among the *Nephelium* species.

PARIH (*Nephelium xerospermoides*)

Parih is a rare specimen from Sarawak and Brunei. The tree grows up to a height of 20 metres in the forests. Leaves are long (17 by 6 centimetres) and elliptical in shape. Fruits are small with reddish-brown, grooved skins.

The thin, sweet, edible flesh surrounds a single, large seed.

The parih is a very vigorous grower and a prolific cropper. Other names used for this fruit include **kalas**, **arut**, and **parapahit**.

MELAJAN (*Nephelium melanomiscum*)

Melajan is also known as **melanjan**, **melanyan** and **melanjau**. It is a rare specimen that is seen growing on a wide variety of soils throughout Sarawak. It also occurs in parts of Peninsular Malaysia. Melajan grows to be a medium-sized tree (18 metres) in the primary or virgin forests. It is a slow growing specimen with smooth bark and a well-rounded crown (unusual in most *Nephelium* species). Leaves are small and pinnate.

The oval-shaped fruits (35 millimetres long) are sparsely covered with long, tapering hairs. The fruit skin is dark red to maroon or purplish in colour and a beautiful shade of purple inside. The sweet, translucent flesh varies in thickness and flavour.

DIMOCARPUS spp.

ISAU (*Dimocarpus longana var. malesiana*)

The isau is native to Sarawak and East Sabah. It is a rare specimen that is occasionally seen growing in the Mixed Dipterocarp Forest on fertile, clay soils. In the 1st Division, Sarawak, the tree was found growing to a height of 18 metres on limestone soils in the low hills. In the 3rd Division a 15 metre high tree was observed growing in the sandy clay of a riverbank. Some isau trees were also seen under cultivation, especially adjacent to the river systems. They were smaller orchard specimens with dense, rounded crowns, usually only growing to 10 metres. Leaves (75 by 25 millimetres) are dark green when

Isau (*Dimocarpus longana var malesiana*)

mature and a beautiful deep red colour in new growth flushes.

Isau fruits are borne in dense clusters on the outside of the tree foliage. They are small (25 millimetres), round and mid-green in colour. Their thin, brittle shells are covered in small bumps and are easily opened. Fruit flesh is translucent, 4 to 6 millimetres thick and surrounds a single, chocolate-brown seed. Isau fruits are very similar to those of the longan. They are sweet and juicy with a delicious musky or melon-like flavour reminiscent of a very sweet watermelon. Other names used for the isau include **menyau**, **merakiang** and **ensiru**.

KAKUS (*Dimocarpus longana var. kakus*)

The kakus is a delicious, 'longan-like' fruit, similar in appearance to the isau, but with a yellow or yellow-brown, pebbly shell. The sweet, musky flesh is a little less juicy than isau with a more distinctive flavour similar to the rockmelon. Fruits are borne in open clusters at the ends of the branches.

The tree is native to Sarawak and East Sabah. The leaves are much larger than those of the isau (some 250 by 100 millimetres or more) and the new foliage is more brown than red in colour. Kakus is a very attractive specimen and crops heavily.

SAU (*Dimocarpus longana var. sau*)

Sau grows naturally in Sarawak. The yellow, pebbly fruits of this tree are reputedly larger and tastier than those of isau or kakus. They grow in compact clusters like those of isau. Tree leaves are large and intermediate in size between those of isau and kakus.

SPINY LONGAN (*Dimocarpus longana var. echinata*)

This rare tree is occasionally found growing in East Sabah. It bears a maroon-red fruit (25 to 32 millimetres) covered in soft, short spines. The edible, translucent flesh is thick, sweet and musk flavoured like the longan.

GURING (*Dimocarpus sp.*)

The guring is native to Sarawak and has small leaves like the isau. It bears a smooth-skinned, brownish-yellow fruit, slightly smaller than isau, with a thin flesh. It is somewhat similar to the cultivated longan, *Euphoria longan*.

BACCAUREA spp.
(*Tampoi*)

There are many species of tampoi found throughout Borneo, Peninsular Malaysia, Java, Sumatra, and the Philippines. Some tampoi names include **ajong**, **larah**, **engkuni**, **puak**, **pugi**, **levan**, **luung keio**, and **umpo**. Several species such as *Baccaurea costulata* are rare and are occasionally seen in the Mixed Dipterocarp Forest in parts of Sarawak. They grow in lowland, clay-rich soils (sometimes in sand or limestone types) or on the lower slopes of the mountains. The tampoi is a small to medium-sized tree with smooth, broad, mid-green leaves. Its pink or white flowers grow in long racemes from the trunk and main branches. Tampoi trees are dioecious, requiring separate male and female trees for fruiting.

All tampoi trees are cold sensitive and should only be grown in warm, frost-free areas. They prefer partial shade for the first few growing years, or permanently. They are slow and difficult to grow outside of their natural forest habitats. Three trees should be grown together.

ORANGE-FLESHED TAMPOI (*Baccaurea costulata*)

These small (50 to 63 millimetres), round fruits hang in profusion on the trunk. Their thick, dull-orange shells split easily when squeezed to reveal six glossy orange segments, arranged in pairs. A small quantity of the sweet, melting flesh is left clinging to each of the six small, flat seeds. The flesh has a tangy juice flavour (perhaps best described as nectarine-apple).

RED-ANGLED TAMPOI, AJONG or UJONG (*Baccaurea angulata*)

This specimen is very attractive, however the fruits are somewhat sour. Its bunches of bright red, five-angled fruits hang from the trunk and larger branches in such profusion as to coat or cover the tree all the way down to ground level.

ENGKUNI (*Baccaurea racemosa*)

The engkuni is another edible tampoi with white fruits (35 millimetres) that ripen to scarlet and are covered in fine, silvery hairs. Other white-fleshed tampois are often sour.

KERANJI
(*Dialium indum and other species*)

The keranji are large forest trees that are commonly cut for timber in East Malaysia. They are native to Malaya, Sarawak and Sabah and are often seen growing near rivers. *Dialium indum* is planted in West Malaysia as a fruit tree. Leaves of a typical keranji tree are dark green, thick and trifoliate (23 by 5 centimetres). Light brown flowers are produced in bunches.

Keranji fruits are small, dull velvety-black in colour with thin, brittle shells. They are easily opened by cracking their eggshell-like pods. These pods help to preserve the 'dried fruit-like' flesh in perfect condition for several months. The fruit is also referred to as the **velvet tamarind**.

The seeds of the keranji are slow to germinate and scarification of the seed coat is recommended. The tree is a slow grower in its natural habitat. It is more vigorous under orchard conditions and may flower in three years from planting out. Trees are semi-deciduous.

KERANJI PAPAN

This small, oblong, thumb-sized fruit (3 centimetres) is slightly flattened and has a bluish-black skin. The rich flesh, similar to a date, surrounds a single, flat seed. This flesh can be sucked or scraped off, somewhat like the flesh of the tamarind, *Tamarindus indica*. It is sweet and dark maroon in colour and rattles loosely in its shell.

KERANJI MADU

Keranji madu means **honey keranji**. It is the sweetest of all the keranji fruits. The fruits are smaller than those of keranji papan, with thicker edible flesh (6 millimetres or more) that clings to the seed. They colour from pale blue to blue-black at maturity. Their flesh is very sweet and caramel-like.

KERANJI AYER or KERANJI AMPLAWAK

This small (2 centimetres) fruit is rounded in shape. The soft, burnt-orange flesh separates easily from the seed. It is more acid in flavour than the other types, like the tamarind. A riverine habitat is ideal.

KERANJI PANJAN

This is a small, sweet fruit like keranji ayer. It is found in the swamp country.

DABAI
(*Canarium odontophyllum*)

Dabai is native to Sarawak and Brunei, growing in altitudes up to 700 metres. It is a very handsome, medium-sized (maximum 21 metres), upright growing tree with large, pinnate leaves. Clusters of olive-like fruits are held above the dark green foliage. Leaves are thin and furry and the twigs are covered with golden down. New foliage emerges like fists of green or red velvet.

Dabai fruits are a startling white colour when immature, turning blue-black when ripe. They are oblong in shape (35 to 40 millimetres long by 20 to 25 millimetres wide) and have a thin, edible skin. The white or yellow flesh inside is 4 to 7 millimetres thick and covers a single, large, three angled seed. The flavour is unique, the texture thick and oily like a good avocado. The fruits are prepared by soaking them in hot water (55°C) for about ten minutes until they soften. They are eaten with a little soy sauce or salt, with a meal or as a savoury snack. They represent a rich energy source with good amounts of oils and proteins. The kernel or nut is also edible.

Dabai trees are dioecious with separate male and female trees required for fruiting. They are very heavy bearers and yields of up to 800 kilogram per tree have been recorded. A planting distance of 9 metres is recommended. Prune back young trees to produce more bushy growth.

There are other related species which are also edible, including the KAMBAYAU, *Dacryodes rostrata*, from Sarawak and Sabah. The tree grows up to a height of 21 metres in the Mixed Dipterocarp Forest and to a maximum altitude of 800 metres.

The fruits are very similar to dabai and are black or blue-black when mature. The flesh is yellowish-purple in colour. A reference from a major upriver Sarawak tribe, the Kayans, reads, 'The fruit is boiled with salt, eaten with rice. It is much wanted'. Some of

Dabai (*Canarium odontophyllum*)

the Iban people consider it superior to dabai. Three forms of the species are recognised in Sarawak. The fruit is also known as **keramoh** and **klamau**.

ENGKALA
(*Litsea garciae*)

A relative of the avocado, this wonderful little savoury fruit comes as a welcome change from the usually sweet tropical fruits. The tree comes from Sarawak and South-west Sabah where it can be seen growing along riverbanks and scattered near villages. The foliage looks somewhat like that of the avocado, however the leaves are longer and droop more gracefully. There have been attempts to cultivate the engkala in Java. In Sabah it is called **pong lobon**.

The round fruits (35 to 45 millimetres in diameter), flattened top and bottom, have thin, edible, bright pink skins. Medium to thick flesh surrounds the single, avocado-like seed (15 to 20 millimetres). It is creamy-white and similar to avocado, but softer with a more delicate flavour. The fruit is prepared by rolling it around a basket, or hitting it with the back of a spoon. The seeds are a source of fat and are used to manufacture candles and soap.

The engkala grows rapidly and tolerates high light levels if well watered.

ENGKABANG
(*Shorea macrophylla*)

The engkabang is native to Borneo where it is also referred to as the **illipe nut**. It is found growing along riverbanks as a large tree (up to 43 metres) with small buttresses and large, pinnate leaves.

The fruit is a large, oval, chocolate coloured nut (maximum length 50 millimetres) with three broad wings. The nut has a good flavour and a high oil content. The oil is extracted and used in cooking like a margarine. It is also used in the manufacture of cosmetics.

The nut is highly prized in the villages whilst the tree bears fruit. The tree grows abundantly in alluvial clay soils on the lower slopes of clay hillsides. It is a rapid grower and is occasionally cultivated.

KUBAL MADU
(*Willoughbea angustifolia*)

Kubal madu is a fruiting liane or vine from Sarawak, Sabah and Brunei, which grows to a length of as much as 90 metres. It grows slowly at first, however, eventually it reaches the canopy layer of even the tallest primary or secondary rainforest. There are copious quantities of latex in all parts of kubal madu, but this is not annoying when eating the fruit. The liane has white flowers and light green, lanceolate leaves. The round fruits (50 to 75 millimetres in diameter) have bright orange skins. The flesh surrounding the large, flat seeds has a sweet, tangy taste, like orange sherbet.

There are five other species of kubal. KUBAL SUSU bears pear-shaped fruit (10 by 6 centimetres) with a somewhat acid flavour. KUBAL AYER and KUBAL ARANG are small and medium-sized fruits (respectively) that are both quite acidic in taste. KUBAL TUSU fruits (to 15 centimetres) resemble kubal susu and have a sweet flavour. KUBAL TABAU bears large, brown fruit with an acid flesh. Kubal ayer is found growing in swamp country and the other species on ridges and mountains to an altitude of 1400 metres. In cultivation, the kubal should be planted adjacent to a trellis and will require pruning.

RANGIL
(*Xanthophyllum sp.*)

The rangil is one of several rare species of *Xanthophyllum* that are seen growing on clay soils in the Mixed Dipterocarp Forests of Sarawak and Sabah. It is a small tree with branches that grow in zig-zag fashion. The small (35 by 25 millimetres), pointed leaves are glossy, dark green in colour.

Rangil's dull surfaced, sperical fruits (5 to 7 centimetres) are borne on the ends of the tree branches. They are green on top and yellow with a tinge of pink below. Their thick shells enclose several pyramid shaped segments of sweet, melting flesh. The fruit shell can be dried and powdered and used as an effective shampoo. There is also a purple variety of rangil. Other names used for the rangil include **langir** and **nyalin**.

MANGIFERA spp.

BAMBANGAN (*Mangifera pajang*)

The bambangan is a rare species in the Mixed Dipterocarp Forests throughout Sarawak and Sabah. It is widely cultivated by the village people, however, and is found growing to an altitude of 1400 metres. Bambangan grows to be a large, majestic tree (to 50 metres) with long, drooping, mango-like leaves (30 by 7 centimetres). Panicles of beautiful, red-purple flowers grow above the foliage.

The large, brown, oval-shaped fruits are between 12.5 and 24 centimetres long. They are eaten fresh, often with savoury or spicy foods as a juicy, acidic, palate refresher. Most fruits are acid tasting, however the best types are sweet enough to eat out of the hand. They are also used in preserving meat.

It is quite possible that the bambangan is better suited to the wet tropics than is the mango, *Mangifera indica*. Other common names used for this fruit include **mawang** and **embang**.

BELUNU (*Mangifera caesia*)

The belunu is also known as **binjai**, **binjai pulut** and **lanyat**. The tree is a rare specimen in the Mixed Dipterocarp Forest in Sarawak and Sabah, however it is cultivated to a limited extent by some local villagers. Belunu also occurs in Indonesia and the Philippines. It is a large, noble tree, growing to 50 metres with thick, lanceolate leaves, held aloft in rosette form. During good years the trees are covered with pink flowers. Belunu fruits arrive in the village markets in September. They are large, potato-shaped fruits with thin, brown skins. They have a very juicy, white flesh which is sweet and unusually fragrant.

Several other *Mangifera* species were observed. The PAHU, *Mangifera sp.*, is a large, green-brown fruit with a sweet taste. The KUINI or WANGI, *Mangifera odorata*, is a strong smelling fruit like a small mango. It is very much in favour with the locals. The tree has beautiful red flowers and deep-purple new foliage. The BAAB, *Mangifera quadrifolia*, is a small, purple-skinned fruit. It is sweet and juicy like a plum. MANGO AYER, *Mangifera longipes*, is similar to the common mango and is frequently seen in East Sabah.

BERANGAN

(*Castanopsis spp.*)

The berangan or chestnut occurs throughout Borneo to an altitude of 1500 metres. They include a range of species of small to medium-sized trees. The fruits are generally small, though one was observed to be as large as 75 by 37 millimetres, with a large kernel. Berangan fruits are said to be eagerly sought after for their edible chestnut or kernel. One species (name not known) is thought to be poisonous.

LANGSATAN

(*Aglaia spp. and Walsura spp.*)

Langsatans are **forest** or **wild langsats** that are found growing throughout Borneo. They are small to medium-sized trees. Several species of each of these genera produce fruits resembling, and comparable with, the langsat. They are mostly slow growing specimens.

MELINJAU

(*Gnetum gnemon var. gnemon*)

The melinjau is also known as **daun sabong** and is native to Indonesia, Malaysia and Africa. It is a shrub or small tree (3 to 5 metres), sometimes cultivated, with oblong to oval-shaped, glossy green leaves (200 by 75 millimetres). In a family of its own, *Gnetum gnemon* is the only gymnosperm with proto-fruits (possibly an evolutionary link between the conifers and the flowering plants or angiosperms). The leaves, flowers and mature fruits are all used as food. The bark yields one of the strongest and best quality fibres known and is commonly used as cordage.

The small, hard, nut-like fruits are roasted, then pounded flat whilst still hot. They are dried in the sun and fried in oil until they puff up into a porous crisp cake known as keropok. Protein content in the leaves, flowers and fruit is about 4-6 per cent.

The tree is often topped to keep it more compact. It can be grown in shade under other trees and tolerates up to three months of drought conditions. Flowering occurs all year round. The seeds are said to be difficult to germinate.

PLAJAU

(*Pentaspadon motleyi*)

The plajau is also known as **uping**, **lakacho** and **plasin**. It is often seen growing along the riverbanks and riverflats throughout Sarawak and Sabah. It is a medium-sized tree with dark green, pinnate leaves. The twigs and new growth are red in colour. It is rather ornamental with a feathery crown and some showy, conspicuous flowers. The edible kernel is shaped like an almond, but more flattened, and is 20 to 25 millimetres long. It has a pleasant taste when fried. There is a mildly irritating property in the sponge-like shells, so care should be taken when opening them.

Plajau is a vigorous specimen that reaches 1.5 metres in the first year of growth.

GARCINIA spp.

(*Kandis*)

Throughout Sarawak and Sabah there are various *Garcinia* species commonly referred to as **kandis** by the locals. They are invariably small, acid fruits with a varying degree of sweetness and a melting, mangosteen-like texture and flavour. The tree is small to medium in size and grows as a lower storey forest specimen, tolerating deep shade. They flower seasonally, usually at night, and emit a powerful smell which has been described as being like 'highly seasoned gravy'. Male and female flowers are borne on separate trees.

Garcinia bancana bears a 7 to 8 centimetre acid fruit. *Garcinia gaudichaudi* is an edible fruit from Sabah. *Garcinia nervosa* is a large Sabah fruit. *Garcinia forbesii* bears fruit with a red skin and a white, good flavoured flesh. *Garcinia hombriana* bears a large, red-skinned, acid tasting fruit.

RECIPES

SAPODILLA ICE-CREAM

2 litres of vanilla ice-cream
1 cup sapodilla

Soften ice cream at room temperature and then blend with ⅔ cup of sapodilla pulp. Mix in the remaining fruit pulp and freeze.

CAIMITO CREAM

4 caimitos
¼ cup sugar
3 oranges
⅔ cup whipped cream

Slice the fruit in half, remove the seeds, and scoop out the pulp. Mix with sugar in a blender. Add the cream and lightly mix. Serve in glasses.

BRAZIL CHERRY JUICE

Rinse fruit and remove the stem and blossom ends. Add just enough water to cover cherries. Stir and mash fruit while simmering gently. When fruit is soft and tender strain the juice through a sieve or cheesecloth.

CASIMIROA SPREAD

1 cup casimiroa
1 cup cream cheese
¼ cup mayonnaise

Mix together in a blender until smooth.

KITEMBILLA FRIES

1 cup kitembilla
3 tablespoons water
2 lightly beaten eggs
2 tablespoons sugar
1 tablespoon baking powder
1 cup flour (sifted)

Sift and mix the dry ingredients together. Add the sugar, water and eggs. Stir and mash in the fruit. Heat some fat and spoon in a little of the mixture at a time, frying until golden brown. Flip over once. Sprinkle with sugar.

CANISTEL MILK SHAKE

¾ cup canistel
1½ cups cold milk
1¾ cups vanilla ice-cream

Mix the canistel and ½ cup of cold milk together in a blender. Add the ice-cream and the remainder of the milk and continue to blend. Serve immediately.

GRUMICHAMA ICE BLOCKS

6 cups grumichama
1¼ cups sugar
½ cup water

Cut the grumichamas in half and remove the seeds. Squeeze all the fruit to a pulp. Boil the water and dissolve sugar, then leave to cool. Place the fruit pulp in the freezing containers and pour in the sugar syrup and freeze.

MANGOSTEEN SORBET

⅔ cup chopped or diced mangosteen
⅔ cup dry champagne
1 egg white
2½ tablespoons sugar
6 lime slices

Peel the fruit, chop or dice the flesh and push through a fine sieve. Stir the champagne into the puree. Whip up the egg white, mix in the sugar, and fold the mixture into the fruit puree and freeze.

LIME AND AVOCADO WHIP

4 avocados
2 limes, squeezed
8 tablespoons cream
4 egg whites
¾ cup icing sugar

Dice avocados and process in a blender. Add the lime juice and the cream and work to a puree. Beat the egg whites until stiff then mix in the icing sugar a little at a time. Fold in the avocado puree and serve.

JABOTICABA ICE-CREAM

3 cups jaboticaba juice
2¾ cups sugar
2 litres milk

To obtain juice, crush fruit in a saucepan and add enough water to just cover the fruit. Boil, and then simmer until tender. Leave overnight. Wash fruit the following day and bring to boil again. Place in a straining bag to drain for about seven hours. Add sugar to juice and stir until dissolved. Stir juice into milk slowly and freeze.

MALAY APPLE DELIGHT

1 large malay apple (peeled)
½ cup water
4 tablespoons sugar
2 egg whites (stiffly beaten)

Dice malay apple and stew with water and sugar until tender soft. Strain, mash, and stir in egg whites. Serve in sherbet glasses.

GUANABANA SHERBET

1½kg ripe guanabana fruit
3 cups water
2⅓ cups sugar

Chop fruit in half lengthwise and remove the core. Place the pulp and seeds in a large bowl. Add 1 cup of water and mash. Strain juice through a sieve, and continue adding the rest of the water and straining thoroughly. Mix in sugar. Freeze in an ice-cream freezer using eight parts ice to one part salt.

AMBARELLA JELLY

ambarellas
water
sugar

Select some firm fruit and slice into several pieces. Just cover with water and bring to boil. Cook for about thirty minutes at low temperature until all the fruit is soft. Strain off the juice and mix with an equal amount of sugar in a deep pan. Bring to boil and continue until setting stage. Pour off into jars and seal.

BREADFRUIT KARNAK

breadfruit, green or unripe
water
1 onion
cheese sauce

Cut breadfruit into small pieces and soak in water for one hour. Drain and place in a casserole dish (uncovered). Pour some cheese sauce with onion over the breadfruit slices. Bake for twenty minutes in a moderate oven.

CARAMBOLA JUICE

Slice fruit into chunks, place in a saucepan, and add just enough water to cover fruit. Cook fruit until soft, tender, and transparent, mashing it along the way. Strain first through a colander, then through several cheesecloth layers. To sweeten, add some sugar. Pour carambola juice over crushed ice in a tall glass. To make carambola punch, add some ginger ale and star shaped carambola slices for decoration.

KIWIFRUIT MOUSSE

500g kiwifruit
¾ cup castor sugar
3 tablespoons water
2 eggs
1 egg yolk
3 teaspoons gelatine, soaked in 3 tablespoons of chilled water
1 cup whipped cream

Place the kiwifruit and half the sugar in the pan with water, cover, and gently cook for about fifteen minutes until soft. Leave to cool, then puree in a blender. Place the eggs, yolk, and the rest of the sugar in a mixing bowl and whisk until thick. Heat to dissolve gelatine, and mix into the kiwifruit puree and cool slightly. Fold this into the egg mixture with the whipped cream. Place in a glass bowl and chill until set. Decorate with piped cream and kiwifruit slices.

CHRISTMAS MORNING SPECIAL

ripe mango
chartreuse
champagne

Slice up some ripe mango and place in glasses. Pour over some chartreuse. Freeze for twenty minutes, then top up with champagne.

GUANABANA REFRESHER

guanabana
soda water
lime juice
crushed ice

Squeeze some juice from some ripe guanabanas and place in a blender or mixer. Add an equal quantity of soda water, some lime juice and crushed ice and mix. Serve up in glasses as a refreshing drink on a hot day.

CARAMBOLA JAM

1¼kg carambola
3½ cups water
750g sugar
juice of 2 lemons
1 teaspoon salt

Cut the winged edges from the carambola fruit and place in a saucepan. Add water and boil until tender (about fifteen minutes). Add the sugar, lemon juice, and salt, and boil until the jam sets. Pour boiling jam into sterilised jars, and seal when cool.

PAPAYA SEED DRESSING

1 cup of cider vinegar
1 tablespoon seasoned salt
1 tablespoon mustard powder
2 tablespoons sugar
2 cups of olive oil
2 tablespoons papaya seeds
1 minced onion

Place sugar, salt, cider vinegar, mustard powder, olive oil, papaya seeds and minced onion in a blender. Blend until smooth and creamy.

BABACO SMOOTHIE

2 babacos
4 bananas
juice of 2 tangelos
1 cup of coconut or macadamia nuts
1-2 tablespoons sugar

Mix together in a blender until smooth. Pour out into glasses, sprinkle with cinnamon and serve immediately.

PAPAYA SORBET

4 cups papaya pulp
juice of 2 lemons
2 egg whites
1⅓ cups of sugar

Mix the papaya pulp, lemon juice, and sugar. Beat the egg whites and fold into the papaya mix. Freeze until firm. Serve in sorbet glasses.

GUAVA JELLY

guavas
juice of 1 lemon
1 cup sugar
water

Slice up guavas and place in a saucepan. Cover with water and boil for half to one hour. Strain mix through cheesecloth and return to boil rapidly until the juice is reduced by one-third to one-half. Add the juice of a lemon while boiling. Add a cup of sugar and boil again. Simmer until ready to set.

UVILLA MOUSSE

18 marshmallows
½ cup water
1 cup uvilla puree
1 cup cream

Heat the marshmallows and water at low heat in a saucepan. Cool. Add uvilla puree and stand until nearly set. Whip cream and add. Freeze.

BELL FRUIT DELIGHT

bell fruit
sugar
water
1-2 drops cochineal
2 teaspoons cornflour

Slice fruit into small pieces, cover with water, sweeten with sugar, and add cochineal. Boil until tender and thicken with cornflour. Serve with cream or ice-cream.

ACEROLA JAM

2kg acerolas
1¼kg sugar
lemon juice

Put fruit in a saucepan, just cover with water and boil. Strain juice, and continue to boil until it is reduced by a third. Add sugar and boil for ten minutes. Bottle.

HOME GROWN COFFEE

One litre (quart) of coffee beans makes about ten cups. Remove the pulp from the seeds, leaving the parchment on. Sun dry the seeds until the parchment is white and brittle, and can be removed with ease. Roast beans and grind ready for brewing.

DURIAN CAKES

1 durian
²/₅ cup sugar

Scoop out the durian flesh from the rind and place in a saucepan. Stir slowly over a medium heat, removing the seeds with a spoon until only the pulp remains. Add the sugar, and stir well. Continue stirring until the mixture coagulates. Remove from heat and shape into cakes.

MACADAMIA NUT TRUFFLES

90g chocolate
1 tablespoon butter
1 egg yolk
1 teaspoon cream
¼ cup roasted macadamia nuts
3 tablespoons drinking chocolate

Melt chocolate in a small bowl over hot water. Add butter and mix. Remove from heat. Add egg yolk and mix, then add cream and mix. Chop macadamia nuts and add to chocolate. Mix well and chill until firm. Shape mixture into twelve balls and roll each in drinking chocolate. Put into paper cases and store in the refrigerator until ready for use.

BLACK PERSIMMON MOUSSE

500g black persimmon pulp
3 heaped tablespoons of full cream milk powder
2 tablespoons of white sugar
2 teaspoons of vanilla essence
2 teaspoons of cocoa powder
⅓ cup milk

Mix all ingredients together and beat well. Serve thoroughly chilled or frozen.

AVOCADO SMOOTHIE

1 large avocado (peeled and seeded)
½ cup cream
2 cups milk
½ teaspoon vanilla essence (optional)
1 teaspoon honey
2 tablespoons castor sugar

Puree the avocado in a blender. Add the cream and blend until smooth. Gradually add milk and vanilla, honey and sugar. Serve in chilled glasses.

CHERIMOYA DACQUIRI

½ cup light rum
2 tablespoons curacao
2 tablespoons lime juice
2 cups crushed ice
1 cup diced cherimoya

Blend all ingredients together. Add sugar or lemonade to taste. Sprinkle nutmeg on top.

DURIAN ICE-CREAM

3-4 pieces of durian
5 cups of coconut cream
⅛ teaspoon salt, plus sugar to taste

Mix all ingredients together. Freeze the mixture, then re-beat.

LYCHEE COCKTAIL

250g lychees
1 cup diced mango
1 cup diced orange
1 tablespoon sugar
2 teaspoons lemon juice

Combine fruits and chill. Add the lemon juice and sugar. Serve up in cocktail glasses.

APPENDIX

SOME SPECIALISED EXOTIC FRUIT NURSERIES

QUEENSLAND

Forest Glen Tropical Fruits, 15 Tarantall Rd, Forest Hill, Qld 4342.

Fitzroy Nursery, PO Box 859, Rockhampton, Qld 4700.

Pioneer Nursery, Eungella Rd, Marian, Qld 4741.

Exotic Groves Nursery, PO Box 125, Innisfail, Qld 4860.

Treefarm (Borneo fruits), El Arish, Qld 4855.

Riversdale Nurseries, PO Box 38, Tully, Qld 4854.

Avondale Nursery, PO Box 30, Smithfield, Qld 4871.

Limberlost Nurseries, PO Freshwater, Qld 4872.

Rainforest Nursery, Reynolds St, Mareeba, Qld 4880.

Rosebud Farm, PO Box 45, Kuranda, Qld 4872.

WESTERN AUSTRALIA

Spreading Chestnut, PO Box 27, Subiaco, WA 6008.

NSW

Green Toes, PO Box 123, North Richmond, NSW 2754.

Rare Fruit Nursery, Lot 41 Nikko Rd, Warnervale, NSW 2259.

Gilbert's Nursery, Pacific Highway, Moorland, NSW 2443

Yarrahapinni Fruit Trees, C/- PO Stuarts Point, NSW 2441

Food Trees, Kalang Rd, Bellingen, NSW 2454.

Vallance's Nursery, Vallance's Rd, Mullumbimby, NSW 2482.

Lychee Plantation of Australia, Lychee Drive, Rosebank, NSW 2480.

Macadamia Plantations of Australia Nursery, Duncan Rd, Dunoon, NSW 2480.

Fruit Spirit Research Nursery, Dorroughby, NSW 2480.

RARE FRUIT GROWER SOCIETIES AND ORGANISATIONS

Rare Fruit Council of Australia Inc., PO Box 707, Cairns, Qld 4870.

Exotic Fruit Growers Association Ltd, PO Box 80, Lismore Heights, NSW 2480.

West Australian Nut and Tree Crop Association, PO Box 565, Subiaco, WA 6008.

Sunshine Coast Subtropical Fruits Association, PO Box 733, Nambour, Qld 4560.

Australian Macadamia Society, PO Box 445, Caboolture, Qld 4510.

Australian Avocado Growers' Federation, PO Box 19, Brisbane Markets, Brisbane Qld 4106.

New Zealand Tree Crops Association, PO Box 1542, Hamilton, New Zealand.

Rare Fruit Council International, 13609 Old Cutler Road, Miami, Florida 33158, USA

California Rare Fruit Growers Inc., The Fullerton Arboretum, California State University, Fullerton, California 92634.

North American Fruit Explorers, C/- Ray Walker, PO Box 711, St Louis, Missouri 63188 USA.

Indoor Citrus and Rare Fruit Society, 176 Coronado Avenue, Los Altos, California 94022 USA.

FURTHER READING

ALEXANDER, D. McE., *Some Avocado Varieties for Australia*, CSIRO (1978).

BALLINGER, R. and J., and SWAAN, H., *Fruit Gardening in South-Eastern Australia*, The Caxton Press, Melbourne (1983).

BAXTER, P., *Growing Fruit in Australia*, Nelson Books, Melbourne (1981).

BURKILL, I. H., *Dictionary of the Economic Products of the Malay Peninsular*, Published by the governments of Malaysia and Singapore by the Ministry of Agriculture and Co-operatives, Kuala Lumpur, (1966).

CHIN, H. F. and YONG, H. S., *Malaysian Fruits in Colour*, Tropical Press, (1980).

CHUA, S. E. and CHUO, S. K., *A Guide to Tropical Fruit Tree Cultivation*, Ministry of National Development, Singapore (1981).

CORONEL, R. E., *Promising Fruits of the Philippines*, University of the Philippines, Laguna, Philippines (1983).

GARNER, R. J., and SAEED AHMED CHAUDHRI Et al., *The Propagation of Tropical Fruit Trees*, Commonwealth Agricultural Bureaux, Kent (1976).

GOULDSTONE, S., *Growing Your Own Food Bearing Plants*, Macmillan, Melbourne (1983).

HEDRICK, U. P., *Sturtevants Edible Plants of the World*, Dover Publications, New York (1972).

HOLTUM, R. E., *Gardening in the Lowlands of Malaysia*, Straits Times Press Ltd, Singapore (1953).

LINDSAY, P., and CULL, B., *Fruit Growing in Warm Climates*, Reed Books, Sydney (1982).

LOTSCHERT, W., and BEESE, G., *Tropical Plants*, Collins, London (1983).

MACMILLAN, H. P., *Tropical Planting and Gardening*, Macmillan and Co., London (1962).

MAXWELL, L. S., *Florida Fruit*, Lewis Maxwell, Tampa, Florida (1967).

MILLER, C. D., BAZORE, K., and BARTOW, M., *Fruits of Hawaii*, The University Press of Hawaii, Honolulu (1965).

MOLESWORTH ALLEN, B., *Common Malaysian Fruits*, Longman, Singapore (1965).

MOLLISON, B., *Permaculture Two*, Tagari Books, Stanley, Tasmania (1984).

POPENOE, W., *Manual of Tropical and Subtropical Fruits*, Hafner Press, New York (1920).

PURSEGLOVE, J. W., *Tropical Crops — Dicotyledons*, Longman, Essex (1968).

RADECKA, H., *The Fruit and Nut Book*, Sphere Books, London (1984).

STEVENSON, L. J., and V., *Fruit for the Home and Garden*, Angus and Robertson Publishers, Sydney (1979).

STURROCK, D., *Fruits for Southern Florida*, Horticultural Books Inc, Stuart, Florida (1959).

VALLANCE, G., *Growing Lychees*, G. Vallance, Mullumbimby, NSW (1982).

WATSON, B. J., LEWIS, W. J., MAGGS, D. H., and PAGE, P. E., *Austrofruit 1*, Queensland Department of Primary Industries, Brisbane (1984).

WILLIAMS, C. N., CHEW, W. Y., and RAJARATNAM, J. A., *Tree and Field Crops of the Wetter Regions of the Tropics*, Longman, Essex (1982).

Tropical Tree Fruits For Australia, Queensland Department of Primary Industries, Brisbane, Qld (1984).

Tree Crops — The 3rd Component, Proceedings of the First Australasian Conference on Tree and Nut Crops, Perth, Western Australia, Cornucopia Press, Perth (1982).

Underexploited Tropical Plants with Promising Economic Value, National Academy of Sciences, Washington (1975).

Home Fruit Growing, New South Wales Department of Agriculture, Sydney (1984).

Rare Fruit Council of Australia Inc Newsletters 1-37 (1980–1986) PO Box 707, Cairns, Qld

BOTANICAL CLASSIFICATIONS

FAMILY	BOTANICAL NAME	COMMON NAMES	ORIGIN
ACTINIDIACEAE	Actinidia deliciosa	*KIWIFRUIT*, Chinese gooseberry, yangtao.	CHINA
ANACARDIACEAE	Bouea macrophylla	*MAPRANG*, kundangan, setar, gandaria.	ASIATIC TROPICS
	Mangifera caesia	*BELUNU*, binjai, binjai pulut, lanyat	ASIATIC TROPICS
	Mangifera indica	*MANGO*, Indian mango, mangga.	INDIA, INDO-CHINA
	Mangifera longipes	*MANGO AYER*.	ASIATIC TROPICS
	Mangifera odorata	*KUINI* or *WANGI*.	ASIATIC TROPICS
	Mangifera pajang	*BAMBANGAN*, mawang, embang.	ASIATIC TROPICS
	Mangifera quadrifolia	*BAAB*.	ASIATIC TROPICS
	Mangifera sp.	*PAHU*.	ASIATIC TROPICS
	Pentaspadon motleyi	*PLAJAU*, uping, lakacho, plasin.	ASIATIC TROPICS
	Spondias cytherea	*AMBARELLA*, Vi-apple, great hog plum, Otaheite apple	POLYNESIA
	Spondias purpurea	*RED MOMBIN*, Spanish plum, purple mombin.	CENTRAL AMERICA
	Spondias tuberosa	*IMBU*.	SOUTH AMERICA
ANNONACEAE	Annona atemoya	*ATEMOYA*, custard apple.	CENTRAL AND SOUTH AMERICA
	Annona cherimola	*CHERIMOYA*, custard apple, cherimoyer.	SOUTH AMERICA
	Annona diversifolia	*ILAMA*, anona blanca, papause, custard apple.	MEXICO AND CENTRAL AMERICA
	Annona muricata	*GUANABANA*, soursop, sirsak, guayabano.	WEST INDIES
	Annona reticulata	*MAMON*, bullocks heart, custard apple.	TROPICAL AMERICA
	Annona squamosa	*SWEETSOP*, sugar apple, custard apple.	CENTRAL AND SOUTH AMERICA
	Rollinia deliciosa	*ROLLINIA*, Amazon custard apple, countess fruit, biriba.	SOUTH AMERICA
	Stelechocarpus burahol	*KEPEL FRUIT*, kepple apple.	INDONESIA
APOCYNACEAE	Carissa grandiflora	*NATAL PLUM*, carissa.	SOUTH AFRICA
	Willoughbea angustifolia	*KUBAL MADU*.	BORNEO
	Willoughbea spp.	*KUBAL SUSU, KUBAL AYER, KUBAL ARANG, KUBAL TUSU, KUBAL TABAU*.	BORNEO

FAMILY	BOTANICAL NAME	COMMON NAMES	ORIGIN
BOMBACACEAE	Durio dulcis	*TUTONG*, red durian, lahong.	BORNEO
	Durio graveolens	*RED-FLESHED DURIAN*, merahan, tabelak.	BORNEO
	Durio kutejensis	*LAI* or *NYEKAK*, ukak, dalit.	BORNEO
	Durio oxleyanus	*BELUDU*, kerantongan.	BORNEO
	Durio zibethinus	*DURIAN*, thurian, civet fruit.	BORNEO
	Durio sp.	*ISU, ISU RAMIN, ISU KUNING.*	BORNEO
	Matisia cordata	*MATISIA*, South American sapote.	SOUTH AMERICA
BURSERACEAE	Canarium odontophyllum	*DABAI.*	BORNEO
CACTACEAE	Hylocereus guatemalensis	*PITAYA*, night blooming cactus.	CENTRAL AND SOUTH AMERICA
CARICACEAE	Carica papaya	*PAPAYA*, papaw, paw paw.	MEXICO AND CENTRAL AMERICA
	Carica pentagona	*BABACO.*	SOUTH AMERICA
CHRYSOBALANACEAE	Chrysobalanus icaco	*ICACO*, cocoplum, guinda.	TROPICAL AND SUBTROPICAL AMERICA
DIPTEROCARPACEAE	Shorea macrophylla	*ENGKABANG*, illipe nut.	BORNEO
EBENACEAE	Diospyros digyna	*BLACK PERSIMMON*, black sapote, chocolate pudding fruit.	MEXICO
	Diospyros discolor	*MABOLO*, velvet apple, butterfruit.	PHILIPPINES
	Diospyros kaki	*PERSIMMON*, Oriental persimmon, kesemek.	CHINA
ELAEOCARPACEAE	Muntingia calabura	*PANAMA BERRY*, capulin, strawberry tree.	CENTRAL AMERICA AND WEST INDIES
EUPHORBIACEAE	Baccaurea angulata	*RED-ANGLED TAMPOI, AJONG* or *UJONG.*	ASIATIC TROPICS
	Baccaurea costulata	*ORANGE-FLESHED TAMPOI.*	ASIATIC TROPICS
	Baccaurea motleyana	*RAMBAI*, menteng negeri.	ASIATIC TROPICS
	Baccaurea racemosa	*ENGKUNI.*	ASIATIC TROPICS
	Blighia sapida	*AKEE*, pan y Quiesto.	WEST AFRICA
FAGACEAE	Castanopsis spp	*BERANGAN*, chestnut.	ASIATIC TROPICS
FLACOURTIACEAE	Dovyalis hebecarpa	*KITEMBILLA*, Ceylon gooseberry, dovyalis.	SRI LANKA
	Flacourtia indica	*RAMONTCHI*, Governor's plum, serali.	ASIATIC TROPICS
GNETACEAE	Gnetum gnemon var. gnemon	*MELINJAU*, daun sabong.	ASIATIC TROPICS, AFRICA
GUTTIFERAE	Garcinia bancana, G. forbesii, G. gaudichaudi, G. hombriana, G. nervosa	*KANDIS.*	BORNEO

FAMILY	BOTANICAL NAME	COMMON NAMES	ORIGIN
GUTTIFERAE	Garcinia livingstonei	IMBE, pama, gupenja.	PORTUGUESE EAST AFRICA
	Garcinia mangostana	MANGOSTEEN, purple mangosteen, manggis.	MALAYSIA, INDONESIA
	Mammea americana	MAMMEA, mamee apple, mammey, mamey.	WEST INDIES, CENTRAL AMERICA
	Rheedia brasiliensis	BAKUPARI, pacura, bacupari.	SOUTH AMERICA
	Rheedia madruno	MADRONO.	SOUTH AMERICA
LAURACEAE	Litsea garciae	ENGKALA, pong lobon.	BORNEO
	Persea americana	AVOCADO, avocado pear, alligator pear, adpukat.	MEXICO, CENTRAL AND SOUTH AMERICA
LEGUMINOSAE	Ceratonia siliqua	CAROB, carob bean, St John's bread.	MEDITERRANEAN
	Dialium indum and other spp.	KERANJI PAPAN, KERANJI MADU, KERANJI AYER, KERANJI PANJAN.	ASIATIC TROPICS
	Inga spp.	INGA, ice-cream bean.	WEST INDIES, CENTRAL AND SOUTH AMERICA
	Tamarindus indica	TAMARIND, tamarindo.	TROPICAL AFRICA
MALPIGHIACEAE	Malpighia glabra	ACEROLA, Barbados cherry, huesito.	WEST INDIES AND CENTRAL AMERICA
MELIACEAE	Aglaia domesticum	LANGSAT/DUKU, lansone, dookoo.	ASIATIC TROPICS
	Aglaia spp. and Walsura spp.	LANGSATAN, wild langsat.	ASIATIC TROPICS
	Sandoricum koetjape	SANTOL, sentul.	ASIATIC TROPICS
MORACEAE	Artocarpus altilis	BREADFRUIT, sukun.	SOUTH-EAST ASIA
	Artocarpus anisophyllus	ENTAWAK, kelidang, tawak, bintau.	ASIATIC TROPICS
	Artocarpus elasticus	TEKALONG, terap, togop, benda, ahbat.	ASIATIC TROPICS
	Artocarpus heterophyllus	JAKFRUIT, nangka, jaca.	INDIA
	Artocarpus integer	CHEMPEDAK, temedak, nakan.	ASIATIC TROPICS
	Artocarpus kemando	PUDAU, pudu, puroh, selibut, chempedak ayer.	ASIATIC TROPICS
	Artocarpus nitidus	SELANKING, butong, empatah, sinojoh.	SOUTH-EAST ASIA
	Artocarpus odoratissimus	MARANG, lemot, tarap.	BORNEO

FAMILY	BOTANICAL NAME	COMMON NAMES	ORIGIN
	Artocarpus sarawakensis	*PINGAN.*	BORNEO
	Artocarpus sericicarpus	*PEDALAI.*	ASIATIC TROPICS
	Pourouma cecropiaefolia	*UVILLA*, Amazon tree grape.	SOUTH AMERICA
MUSACEAE	Musa spp.	*BANANA.*	SOUTH-EAST ASIA
MYRISTICACEAE	Myristica fragrans	*NUTMEG*, pala.	MOLUCCAS
MYRTACEAE	Eugenia brasiliensis	*GRUMICHAMA*, Brazil cherry.	SOUTH AMERICA
	Eugenia luschnathiana	*PITOMBA.*	SOUTH AMERICA
	Eugenia uniflora	*BRAZIL CHERRY*, Surinam cherry, pitanga.	SOUTH AMERICA
	Feijoa sellowiana	*FEIJOA*, pineapple guava, guavasteen.	SOUTH AMERICA
	Myrciaria cauliflora	*JABOTICABA.*	SOUTH AMERICA
	Myrciaria paraensis	*CAMU CAMU.*	SOUTH AMERICA
	Psidium guajava	*GUAVA*, guayaba, apple guava, jambu batu.	AMERICAN TROPICS
	Syzygium aqueum	*BELL FRUIT*, water apple, water cherry, watery rose-apple.	INDIA
	Syzygium cumini	*JAMBOLAN*, Jambolan plum, Malabar plum, Java plum, duhat.	INDIA, BURMA, CEYLON
	Syzygium jambos	*ROSEAPPLE*, jambu mawar.	INDONESIA MALAYSIA
	Syzygium malaccense	*MALAY APPLE*, jambu merah, Malay roseapple, ohia, pomerac.	MALAYSIA
	Syzygium samarangense	*WAX JAMBU*, Java roseapple, Samarang roseapple.	MALAYSIA, ANDAMAN IS.,
OXALIDACEAE	Averrhoa bilimbi	*BILIMBI*, cucumber tree.	MALAYSIA
	Averrhoa carambola	*CARAMBOLA*, five corners, starfruit.	MALAYSIA, INDONESIA
PALMAE	Butia capitata	*JELLY PALM*, butia, wine palm.	SOUTH AMERICA
	Cocos nucifera	*COCONUT.*	MELANESIA
	Bactris gasipaes	*PEJIBAYE*, peach palm, chanto duro.	SOUTH AMERICA
	Phoenix dactylifera	*DATE*, date palm.	PERSIAN GULF REGION
	Salacca edulis	*SALAK*, salak palm, salac.	ASIATIC TROPICS
PASSIFLORACEAE	Passiflora edulis	*PURPLE PASSIONFRUIT.*	SOUTH AMERICA
POLYGALACEAE	Xanthophyllum sp.	*RANGIL*, langir, nyalin.	BORNEO
PROTEACEAE	Macadamia integrifolia/tetraphylla	*MACADAMIA*, Qld bush nut, bauple nut.	AUSTRALIA

FAMILY	BOTANICAL NAME	COMMON NAMES	ORIGIN
PUNICACEAE	Punica granatum	*POMEGRANATE*, delima	IRAN
RHAMNACEAE	Hovenia dulcis	*CHINESE RAISIN TREE*, Japanese raisin tree	CHINA
	Ziziphus jujuba	*CHINESE JUJUBE*, Chinese date.	CHINA
	Ziziphus mauritiana	*INDIAN JUJUBE*, ber, bedara, manzanitas.	INDIA
ROSACEAE	Eriobotrya japonica	*LOQUAT.*	CHINA
	Prunus serotina var. salicifolia	*CAPULIN*, Capulin cherry.	CENTRAL AND SOUTH AMERICA
	Pyrus pyrifolia	*ASIAN PEAR*, nashifruit, Chinese pear.	ASIA
RUBIACEAE	Coffea arabica	*ARABICA COFFEE*, coffee.	AFRICA
RUTACEAE	Aegle marmelos	*BAEL FRUIT*, bel fruit.	INDIA
	Casimiroa edulis	*CASIMIROA*, white sapote, cochiztzapotl, zapote blanco.	MEXICO AND CENTRAL AMERICA
	Casimiroa tetrameria	*MATASANO*, woolly leaf white sapote, yellow sapote, Guatemalan sapote.	MEXICO AND CENTRAL AMERICA
	Citrus aurantifolia/ latifolia	*LIME*, West Indian lime, Tahiti lime.	ASIA
	Citrus maxima	*PUMMELO*, shaddock, pamplemouse.	MALAYSIA AND THAILAND
	Clausena lansium	*WAMPI*, wampee	CHINA
SAPINDACEAE	Dimocarpus longana var. echinata	*SPINY LONGAN.*	BORNEO
	Dimocarpus longana var. kakus	*KAKUS.*	BORNEO
	Dimocarpus longana var. malesiana	*ISAU*, menyau, ensiru.	BORNEO
	Dimocarpus longana var. sau	*SAU.*	BORNEO
	Dimocarpus sp.	*GURING.*	BORNEO
	Euphoria longan	*LONGAN*, lengkeng, lungan.	CHINA OR INDIA
	Litchi chinensis	*LYCHEE*, litchi, litchee.	CHINA
	Melicoccus bijugatus	*MAMONCILLO*, Spanish lime, genip, honeyberry	CENTRAL AMERICA
	Nephelium lappaceum	*RAMBUTAN*, litchi cheval, rambutao, ngo phruan.	MALAYSIA AND SUMATRA
	Nephelium maingayi	*LAIT.*	BORNEO
	Nephelium melanomiscum	*MELAJAN*, melanjan.	MALAYSIA AND INDONESIA
	Nephelium mutabile	*MERITAM* or *PULASAN*, maha, ma.	MALAYSIA AND INDONESIA
	Nephelium xerospermoides	*PARIH*, kalas, arut.	BORNEO
	Nephelium sp.	*PANGKAL.*	BORNEO
	Nephelium sp.	*SIBAU*, rupah.	BORNEO

FAMILY	BOTANICAL NAME	COMMON NAMES	ORIGIN
	Nephelium sp.	*TITIDAN.*	BORNEO
	Pometia pinnata	*TAUN or DAWA*, Fiji longan.	PAPUA NEW GUINEA, FIJI, SAMOA, SOLOMON ISLANDS
SAPOTACEAE	Chrysophyllum cainito	*CAIMITO*, star apple, sawo duren.	WEST INDIES, CENTRAL AMERICA
	Manilkara zapota	*SAPODILLA*, chiko, chiku, chico, nispero, naseberry.	CENTRAL AMERICA
	Pouteria caimito	*ABIU*, caimo, cauje.	SOUTH AMERICA
	Pouteria campechiana	*CANISTEL*, yellow sapote, egg fruit, tiesa, huevo vegetal.	MEXICO AND CENTRAL AMERICA
	Pouteria obovata	*LUCMO*, lucuma, lucma, tiesa de peru.	SOUTH AMERICA
	Pouteria sapota	*MAMEY SAPOTE*, mamey, mammee, zabotillo, mamey colorado.	CENTRAL AMERICA
	Pouteria viride	*GREEN SAPOTE*, injerto.	CENTRAL AMERICA
	Synsepalum dulcificum	*MIRACLE FRUIT*, miraculous berry.	TROPICAL WEST AFRICA
SOLANACEAE	Cyphomandra betacea	*TAMARILLO*, tree tomato, terong belanda.	SOUTH AMERICA
	Solanum hyporhodium	*COCONA*, coco.	SOUTH AMERICA
	Solanum muricatum	*PEPINO*, papino, melon pear.	SOUTH AMERICA
	Solanum quitoense	*NARANJILLA*, lulu.	SOUTH AMERICA

INDEX